MYCOFACTORIES

Editor

Ana Lúcia Leitão

eBooks End User License Agreement

CONTENT

FOREWORD

Among the estimated 1.5 million species of fungi, about 100,000 have so far been described, and therefore, concerted efforts are globally needed to fully understand their diversity and exploitation. Fungi are considered to be the second largest group of organisms after insects, and they have been extensively studied because of their fascinating nature, enormous capability to cope with and survive in a great variety of environments on Earth, and yield products useful in medicine, nutrition and industry.

Fungal enzymes have been extensively used in food industry for several decades as processing aids since they have several attributes that make them suitable for this purpose. They are being used in baking, cheese manufacture, fruit juice and wine making, brewery and starch saccharification and as animal feed supplements to mention a few. Fungal enzymes in general are non-toxic and speed up chemical reactions with utmost specificity at ambient and low temperatures, pressure and neutral pH. A large industry, therefore, exists to serve this need all over the world. Rapid strides are being made in discovering new fungal sources of useful enzymes, their cloning and over expression and improving their useful properties by enzyme engineering. Lignocellulolytic fungi and their enzymes are being investigated for their potential use in producing wealth from wastes and bioconversion of abundantly available and renewable agricultural and forest residues to valuable products such as xylooligosaccharies, bioethanol and SCP. Several fungi have recently been shown to be useful in biofiltration of waste gases and thus aid in mitigating air pollution.

Penicillium camemberti and *P. roqueforti* have been used for long time in ripening of cheese because their mycelium contributes significantly to the product flavour, regulates moisture loss, and prevents the development of potentially mycotoxigenic fungal species. The fermented products made employing *Monascus* are vastly consumed as a popular dietary supplement that claims to prevent or ameliorate hypertension, hypercholesterolemia, and hyperlipidemia. The metabolites of *Monascus* contain bioactive ingredients such as ankaflavin, monacolins, γ-aminobutyric acid, and dimerumic acid, and among these, monacolin K is well known because this is considered to aid in controlling serum cholesterol level. Research efforts are underway to elucidate molecular mechanisms of metabolite synthesis, optimizing cultivation conditions for maximizing beneficial ingredients and to minimize toxicity.

Moulds such as *Aspergillus oryzae, A. niger, Fusarium graminearum, Trichoderma viride* and others are being used for heterologous gene expression because they are GRAS organisms, secrete a wide range of proteins, glycosylation similar to higher eukaryotes, integration of introduced DNA into fungal genome, thus avoiding the need to maintain selection pressure and strong promoters.

In the book entitled 'Mycofactories' a conscious effort has been made to choose contributors who have extensive experience and expertise on the topics stated above. The book is aimed at giving the readers an overview of recent developments in the use of fungi for a variety of purposes, product and process development, in-depth understanding of the underlying mechanisms, and traditional and novel applications. The book will be very useful as a ready reference for students and researchers in the diverse areas of biology, biotechnology and environmental sciences.

T. Satyanarayana
Department of Microbiology
University of Delhi South campus
New Delhi – 110021, India

PREFACE

Many fungi live saprophytically in the soil having an important role in nature in carbon recycling. They are well known for their ability to adapt to environments of high osmolarity by a mechanism involving the accumulation of low-molecular weight compounds that maintain positive turgor pressure. Indeed, fungi produce a diverse array of secondary metabolites. These molecules, that are not necessary for normal cell growth, have a tremendous impact on industry and society. As a biotechnology tools, fungi have the advantage of being relatively easy to growth on fermenters being suited for large scale production. They have been successfully employed for biotransformations ranging from food manufacturing to drug design.

The production of penicillin by filamentous fungi has been pointed out as the beginning of modern pharmaceutical industry. Recently, the synthesis of naturally occurring bioactive microbial metabolites like pyripyropenes, arisugacins in order to create novel medicines for specific clinical conditions such as atherosclerosis, Alzheimer's disease, cancer, inflammation, and osteoporosis, among others, opened a new perspective for fungi applications. On the other hand, these microorganisms are responsible to secrete high levels of extracellular proteins such as plant cell wall hydrolyzing enzymes employed for the hydrolysis of lignocellulosic materials. Furthermore, they are known to produce a series of hydrocarbon derivatives normally present in diesel fuel. The use of fungi on bioremediation field of environmental hazardous compounds to reduce levels of heavy metals, polycyclic aromatic hydrocarbons (PAHs), phenol and phenolic compounds is clearly in expansion. Fungi are considered as ideal green catalysts having great biotechnological impact due to their limited nutritional requirements, and their broad substrate specificity. This very high metabolic diversity has been actively exploited for many years and will be developed in the future, as new mankind challenge appeared.

Highlights included in the e-book are focused on the current and future trends about the applications of fungi as a main source for the production of enzymes and for the manufacturing of food derivatives, applications in the bioremediation field, production of pigments and other food additives.

Human needs and challenges for designing novel strategies to develop more efficient cell factories made this book interesting and actual. This e-Book publishes a substantial share of the most significant current research in the area of fungal biotechnology, and constitutes an effort to compile high level research in biotechnology with fungi as main tools.

Ana Lúcia Leitão

Faculdade de Ciências e Tecnologia,
Universidade Nova de Lisboa,
Campus de Caparica,
2829-516 Caparica, Portugal

LIST OF CONTRIBUTORS

G.D.Y. Boorgula

Departmental of Microbiology
Kakatiya University
Warangal
Andhra Pradesh-506009, India

Renato Chávez

Departamento de Biología
Facultad de Química y Biología
Universidad de Santiago de Chile (USACH)
Santiago, Chile

Francisco J. Enguita

Unidade de Biologia Celular
Instituto de Medicina Molecular
Faculdade de Medicina
Universidade de Lisboa
Av. Prof. Egas Moniz
1649-028 Lisboa, Portugal

Francisco Fierro

Departamento de Biotecnología
División de Ciencias Biológicas y de la Salud
Universidad Autónoma Metropolitana-Iztapalala
Mexico D.F., Mexico

Jean Marie François

Université de Toulouse
INSA, UPS, INP
135 Avenue de Rangueil
F-31077 Toulouse, France

Ramón O. García-Rico

Departamento de Microbiología
Facultad de Ciencias Básicas
Universidad de Pamplona
Pamplona, Colombia

Olivier Guais

Université de Toulouse
INSA, UPS, INP
135 Avenue de Rangueil
F-31077 Toulouse, France

Sergio Hernández

Departamento de Ingeniería de Procesos e Hidráulica
Universidad Autónoma Metropolitana-Iztapalapa
Apdo. Postal 55-534
CP 09340, Mexico D.F., Mexico

Federico Laich

Unidad de Microbiologia Aplicada
Instituto Canario de Investigaciones Agrarias
Santa Cruz de Tenerife, Spain

Min-Hsiung Lee

Department of Agricultural Chemistry
National Taiwan University
Taiwan

Ana Lúcia Leitão

Faculdade de Ciências e Tecnologia,
Universidade Nova de Lisboa
Unidade de Biotecnologia Ambiental
Campus de Caparica
2829-516 Caparica, Portugal

Yii-Lih Lin

Department of Agricultural Chemistry
National Taiwan University
Taiwan

Aurelio Moraleda-Muñoz

Departamento de Microbiología
Facultad de Ciencias
Instituto de Biotecnología
Universidad de Granada
Av. De Fuentenueva s/n
E-18071 Granada, Spain

Juana Pérez

Departamento de Microbiología
Facultad de Ciencias
Instituto de Biotecnología
Universidad de Granada
Av. De Fuentenueva s/n
E-18071 Granada, Spain

J.L. Uma Maheswar Rao

Faculdade de Ciências e Tecnologia,
Universidade Nova de Lisboa
Unidade de Biotecnologia Ambiental
Campus de Caparica
2829-516 Caparica, Portugal

Sergio Revah

Departamento de Procesos y Tecnología
Universidad Autónoma Metroplitana Cuajimalpa
c/o IPH, UAM-Iztapalapa
Av. San Rafael Atlixco No. 186
09340 Mexico D.F., Mexico

Alberto Vergara-Fernández

Escuela de Ingeniería Ambiental
Facultad de Ingeniería
Universidad Católica de Temuco
Rudecindo Ortega 02950
Casilla 15-D
Temuco, Chile

Nan-Wei Su

Department of Agricultural Chemistry
National Taiwan University
Taiwan

Teng-Hsu Wang

Department of Agricultural Chemistry
National Taiwan University
Taiwan

CHAPTER 1

Fungal Enzymes: Present Scenario and Future Perspectives

J.L. Uma Maheswar Rao[1]*, G.D.Y. Boorgula[2] and Ana Lúcia Leitão[1]

[1]*Faculdade de Ciências e Tecnologia, Universidade Nova de Lisboa, Unidade de Biotecnologia Ambiental, Campus de Caparica, 2829-516 Caparica, Portugal and* [2]*Departamental of Microbiology, Kakatiya University, Warangal, Andhra Pradesh-506009, Índia.*

Abstract: As a conservative estimate, some 120,000 species of fungi have been isolated. Many of these have been screened for their ability to produce industrially sound products. Fungi have been important in both ancient and modern biotechnological processes. They are of excellent value in nutrition, processes and products that utilize fungi include production of sugars, antibiotics, enzymes, organic acids, baking, brewing, alcohols, and numerous pharmaceuticals. α-Amylases and glucoamylases are used in the conversion of starch into different sugar syrups. Industrial applications generally require amylases with a very specific hydrolysis profile. The commercially used fungal amylases have certain limitations such as moderate thermostability, acidic pH requirement, and slow catalytic activity that increase the process cost. The importance of retrogradation of starch fraction in bread staling has been emphasized. A loss of more than US$1 billion is incurred in USA alone every year due to the staling of bread in winters. There is a need for good additives and enzymes for preventing staling and to improve the texture and shelf life of baked products. Pectinases are one of the upcoming enzymes of fruit and textile industries. These enzymes are required for the break down of complex polysaccharides of plant tissues into simpler molecules like galacturonic acids. The role of acidic pectinases in bringing down the cloudiness and bitterness of fruit juices is well established. The production of pectinolytic enzymes has been widely explored in filamentous fungi. However, there are a very few reports on the production of pectinases by thermophilic moulds for food applications. About two thirds of the phosphorus in plant ingredients for pigs and poultry is in the form of salts of phytic acid (myoinositol hexakisphosphates, phytates), which are not very soluble and of very limited digestibility. This area of research has advanced to the extent that enzymes are commonly required in poultry diets to enhance the nutritive value of cereals. This review focuses attention on the present status of knowledge on the production, characterization, and potential applications of fungal biocatalysts (alpha amylases, glucoamylases, pectinases and phytases) in food industry.

INTRODUCTION

Biopolymers such as cellulose, starch, pectin, xylan and lignin are abundant in nature and constitute a major part of living matter. After cellulose, starch is the second most abundant polysaccharide produced by plants. The ability to utilize these substances as carbon and energy sources is widely distributed among animals, plants and microorganisms. A wide variety of bacteria, fungi and yeasts produce a large array of extracellular enzymes to degrade these substances in different environmental niches [1].

The great diversity of polysaccharide-hydrolyzing enzymes have fulfilled the requirement of enzyme industry over the last few decades and also led to interest in further screening programmes to look for enzymes with novel properties.

Plants synthesize starch as a result of photosynthesis, the process during which energy from sunlight is converted into chemical energy. Starch is synthesized in plastids found in leaves as a storage compound. It is also synthesized in amyloplasts found in tubers, seeds, and roots as a long-term storage compound. Latter large amounts of starch accumulate as water-insoluble granules.

The shape and diameter of these granules depend on their botanical origin. For commercially important starch sources, the granule sizes range from 1.0 (maize starch) to 5.0 μm (potato starch) [2]. Starch is a polymer of glucose linked to one another through the C1 oxygen, known as the glycosidic bond. The glycosidic bond is stable at high pH but hydrolyzed at low pH. At the end of the polymeric chain, a latent aldehyde group is present. This group is known as the reducing end.

Address correspondence to J.L. Uma Maheswar Rao: Faculdade de Ciências e Tecnologia, Universidade Nova de Lisboa, Unidade de Biotecnologia Ambiental, Campus de Caparica, 2829-516 Caparica, Portugal; E-mail: dear_mahesh@rediffmail.com

Natural starch is insoluble in water, and is present in plant cells in the form of microscopic granules. The size and shape of starch granules are characteristic of the plant. Heating in water weakens the hydrogen bonds that bind the granules and causes swelling and gelatinization. This macromolecule is composed of two high molecular weight components: amylose (15-20%) and amylopectin (75-85%). Amylose consists of a linear chain of glucose residues linked by α-1,4- glycosidic bonds. It is twisted into a helical coil with about six glucose residues per turn. It complexes with iodine and forms an intense blue color complex. Amylopectin, on the other hand, is a branched polymer containing α-1,4 linked glucan with α-1,6 linked branch points at every 17-26 glucose residues. Unlike amylose, amylopectin is relatively stable in aqueous solutions and does not form compact aggregates. Due to the branched structure, iodine binding power is less pronounced and this gives a red-violet color. Amylose and amylopectin show different physical properties due to difference in their structure. Their ratios vary in different natural starches. The waxy starches contain no amylose whereas amylomaize starch contains upto 80% of amylose. In general, amylose and amylopectin ratios are in the range between 1:3 and 1:4. These variations in structure and composition of native starches give rise to differences in their degradability by various amylolytic enzymes.

Starch is the storage polysaccharide whereas cellulose is the structural component of plants. Both polysaccharides are composed exclusively of glucose units and are bound by glycosidic linkages. Starch is the principal food reserve in the plant kingdom and serves as a bulk nutrient and energy source in animal and human diet. The glucose residues in starch are linked in the alpha orientation *i.e.* α-glucan. Due to the different anomeric conformations of these polymers, different enzymatic systems are required for their degradation. Glycogen is the counterpart of starch in animals. Starch can be extracted commercially from many raw materials including maize, wheat, barley, potato, rice, oat, cassava, sorghum and others [1, 3]. Starch, a major component of agricultural crops, is an important substrate for chemical and enzyme production because its higher density facilitates prolonged storage and decreased transportation and pre- treatment costs [4].

A large-scale starch processing industry has emerged in the last century. In the past decades, we have seen a shift from the acid hydrolysis of starch to the use of starch-hydrolyzing enzymes in the production of maltodextrin, modified starches, and maltose, glucose and fructose syrups. Amylases constitute about 15% of the world's enzyme sales [5]. Besides the use in starch hydrolysis, starch-hydrolyzing enzymes are also used in a number of other industrial applications, such as laundry and porcelain detergents and as an anti-staling agent in baking [5, 6].

Starch as a Substrate

Starch is a glucose polymer and is one of the most widely available plant polysaccharides. As a chemical substance, starch (amylum) is by no means uniform, but a mixture of polysaccharides, which in one sense are remarkably uniform in composition, being polymers of glycosidically linked units of α-D-glucopyranose in the 4C1 conformation.

Although starch consists of α-D-glucosyl units joined by the α-1,4 or α-1,6 linkages, the molecular structures of each major constituent (amylose and amylopectin) defy precise definition so much, that differences in composition, morphology and properties, and therefore 'starch' is not an entity. The chain lengths of the essentially linear component (amylose) are variable, and there are side chains resulting from the action of branching enzymes. The relative arrangement of amylose and amylopectin within the well-ordered granular structure of starch is still a subject of much research owing to technical importance of starch, and the fundamental need to correlate the gelatinization and solution processes with physical properties, especially rheological behavior [7].

Molecules comprise big glucopyranose units linked by α-1, 4 form hollow helices. These may twist around each other in double helices or pack closely together without intertwining, depending on the relative extension of the helices. Amylose and amylopectin molecules pack together in granules to form concentrated storage deposits. Since starch is a polysaccharide, it has to be relatively rapidly degradable. Its hydrolysis is facilitated by the highly branched nature of amylopectin, with relatively short branches, and the helical nature of the longer amylose molecules, which do not pack as tightly together as cellulose and chitin molecules [8].

Amylose and Amylopectin

The amylose fraction assumes to exist in the granule as an entity that is largely separated from the amylopectin fraction. Thus amylose is able to leach out of the granules, melt and solubilize. Amylose is stable in solution at temperatures greater than 60-70 °C, but it aggregates by association or crystallization on cooling, and forms gels or precipitates almost instantaneously. Once such a gel or precipitate is formed, a temperature greater than 120 °C is required to dissolve the amylose. The larger branched amylopectin forms viscous solutions, which are stable in water at room temperature for a few days. After a longer storage at lower temperatures, amylopectin can also associate to form weak gels [6].

Gelatinization

Heating of a starch suspension above a critical temperature results in a multistage process called gelatinization, which involves tangential swelling of the amorphous regions of the granule, disruption of the readily ordered structures, and eventually opening of the crystal structures as the polymer chain becomes increasingly hydrated. Gelatinization has been defined as phase transition of the starch granules from an ordered state, which takes place during heating in excess water. As long as water is abundant, gelatinization will occur at a fixed range, normally 60-70 °C. The swollen granules are enriched in amylopectin. The linear amylose diffuses out of the swollen granule during and after gelatinization, and makes up the continuous gel phase outside the granules. Swelling of granules into gel particles results in an increase in viscosity. Gelatinization greatly enhances the chemical reactivity of inert starch granules towards amylolytic enzymes; this has become the chosen method of enhancing starch hydrolysis, and has been widely adopted in the manufacture of starch syrups [6, 9].

Industries using enzymes on a large scale include brewing, baking, detergents, dairy, starch processing, leather and textiles. Approximately 80% of enzymes produced by fermentation and sold on an industrial scale are hydrolytic, and almost all are extracellular [10].

Starch - Hydrolyzing Enzymes

Amylolytic enzymes (α-glucanases) hydrolyze the glycosidic linkages in various α-glucans.

The action of these enzymes can be divided into two categories: endoamylases and exoamylases. **Endoamylases**: The dextrinogenic or liquefying amylases act randomly on the α–1,4 linkages only. Their action results in the formation of linear and branched oligosaccharides of varying chain lengths (dextrins). The α–amylases are a type of endoamylases. **Exoamylases**: The saccharifying or saccharogenic amylases hydrolyze polysaccharides from the non-reducing end successively, resulting in short end products.

One type cleaves each bond to produce solely β-glucose (glucoamylases), and another type breaks every alternate bond to produce maltose (β-amylases) (Fig. **1**).

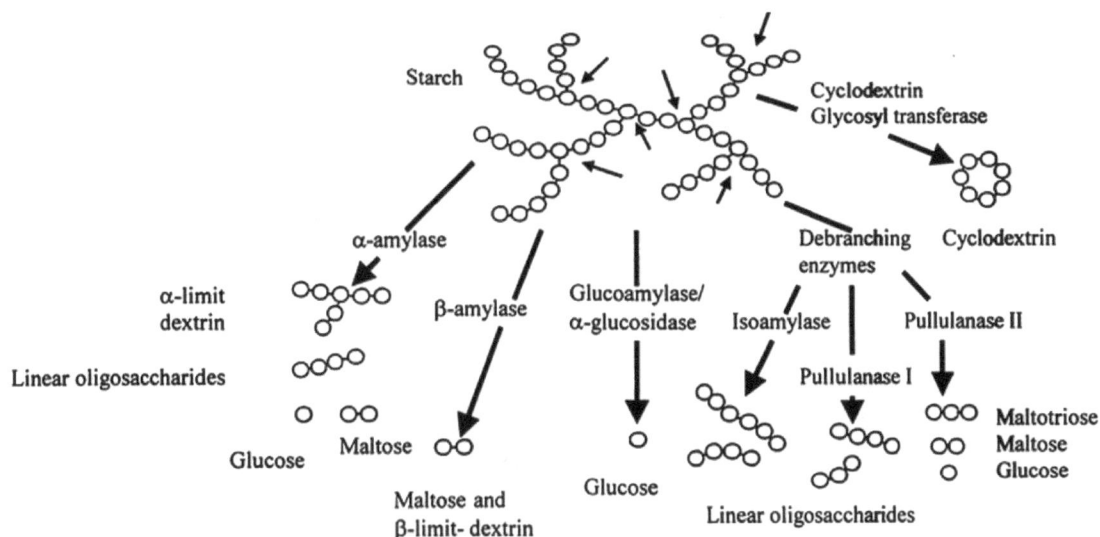

Figure 1: Action points of various hydrolases on starch and their products. Small arrows show the branching points [1].

α-Amylases (1,4-α-D-Glucan Glucanohydrolase, E.C.3.2.1.1.):

α-Amylases are extracellular enzymes, which hydrolyze α-1,4 glycosidic linkages of starch randomly and are capable of bypassing the branching points. They are widely distributed among microbes, which liberate linear and branched oligosaccharides of varying lengths as well as glucose from starch. Such branched oligosaccharides are also called α-limit dextrins. The end products have an α-configuration at C_1 [1]. The action of α-amylases is accompanied by a rapid loss of viscosity and a rapid reduction in iodine staining power due to the formation of oligosaccharides. α-Amylases are divided into two categories according to degree of hydrolysis of substrate, saccharifying α-amylases, which produces free sugars and liquefying α-amylases that break down the starch polymer but do not produce free sugars. Liquefying α-amylases cause more rapid reduction in viscosity of starch pastes, whereas the saccharifying α-amylases reduce the viscosity less rapidly in comparison with the amount of reducing sugars released.

Commercial applications of α-amylases are numerous. Probably the largest volume is used for thinning of starch in the liquefaction process in the sugar, alcohol and brewing industries. They are also used in desizing of fabrics, in the baking industry, production of adhesives, pharmaceuticals, detergents, sewage treatment and animal feeds.

β-Amylases (1, 4-α–D-Glucan Maltohydrolase, E.C.3.2.1.2):

β-Amylases are usually of plant origin, but some microbial producers are also known. This exoenzyme hydolyzes the alternate α-1,4 linkage from the non-reducing end, causing inversion of the anomeric configuration of the liberated maltose to its β-form, thus the name of β-amylase. This enzyme is incapable of bypassing α–1,6-linkages of branched substrates such as amylopectin. The degradation of such substrates is therefore, incomplete. Consequently the end products are maltose and high molecular weight β-limit dextrins.

In some cases, microorganisms form similar exoacting enzymes, but instead of maltose, other oligosaccharides with defined size (maltotetraose, maltohexose etc.) are formed, the reducing ends of which have α-configuration. Food and beverage industries employ β-amylase to convert starch into maltose [11].

Glucoamylase (1,4-α-D-Glucan Glucanohydrolase, E.C.3.2.1.3):

It attacks α–1,4 linkages of α-glucans from the non-reducing ends and also α–1,6 linkages, but at a very slow pace. It preferentially degrades polysaccharides with a high molecular weight. These enzymes hydrolyze starch to give glucose in theoretically 100% yield [12]. Glucoamylase has got wide applications in the food and fermentation industries; in the processes of starch saccharification, brewing and distillation [13]. It has got 14% distribution and sale (second position) on worldwide distribution and sales of industrial enzymes [12]. Glucoamylase, also known as amyloglucosidase and γ–amylase, is a typical fungal enzyme. The reaction rate decreases with the decreasing chain length of the dextrin substrate, and maltose is attacked most slowly [14]. Prolonged incubation time and high concentration of starch usually cause reversion of the reaction catalyzed by glucoamylase. This results in the formation of maltose, isomaltose and some oligosaccharides [1]. Glucoamylases are commercially important in the hydrolysis of starch for the production of glucose and fructose syrups for use in confectionary and ethanol production [11, 15].

Cyclodextrin Glycosyl Transferase (CGTase) {1,4-α-D-Glucan 4-α- D -(1,4 α-D-Glucano)- Transferase (Cyclizing), E.C.2.4.1.19}:

CGTase, an enzyme found only in bacteria, produces a series of α, β, γ cyclodextrins (rings made up of 6,7 and 8 glucose units respectively, bound by α–1,4 bonds) from starch, amylose and other polysaccharides. In addition, CGTases also catalyze coupling reaction by which the rings are opened and transferred to co-substrates like glucose, maltose or sucrose. The enzyme also catalyzes disproportionation reaction in which one or more glycosyl units are transferred between linear oligosaccharides. Finally, a hydrolysis reaction is catalyzed by the enzymes producing dextrins.

α-Glucosidase (α-D-Glucoside Glucohydrolase, E.C.3.2.1.20):

α–Glucosidase hydrolyzes α–1,4-and/or α–1,6-linkages of saccharides, that are formed by the action of other amylolytic enzymes and liberates α–D glucose units from the non-reducing end. This appears to be the final enzyme involved in the breakdown of starch. Most enzymes show high affinity towards maltose and so they are referred to as maltases.

Isoamylase (Glycogen 6-Glucanohydrolase, E.C.3.2.1.68):

Isoamylase hydrolyses the α–1,6-linkages in polysaccharides such as amylopectin, glycogen and branched dextrins in an exo-fashion. However, it cannot degrade α–1,6 linkages of pullulan. Isoamylases are used to debranch starch in the production of glucose and maltose syrups [11].

Pullulanase:

Pullulanases are the debranching enzymes that hydrolyze α–1,6- linkages of amylopectin and pullulan, but their activity on glycogen is very weak.

Pullulanase type I (α-dextrin 6 – glucanohydrolase, E.C.3.2.1.4)

It is a typical bacterial enzyme that is specific for α–1,6 linkages in amylopectin or its degradation products and is unable to attack α–1,4- linkages in α–glucans. It attacks exclusively the α–1,6-linkages in pullulan in a random endofashion and causes its complete conversion to maltotriose.

Pullulanase type II (amylase–pullulanase/amylopullulanase)

An enzyme distributed mostly among anaerobic bacteria, and is capable of hydrolyzing α-1,4 linkages in addition to the branch points (α-1,6 linkages), causing the complete conversion of starch to small sugars without the requirement for other enzymes.

Conventional Starch Hydrolysis Process

The main amylolytic (starch hydrolyzing) enzymes used in the starch industry for the production of glucose, maltose and maltooligosaccharides are α-amylase, β-amylase, glucoamylase and pullulanase [4]. The properties of enzymes used in the process determine the conditions under which the starch processing must operate. Glucose and maltose syrups are usually produced from starch in a two step process-liquefaction and saccharification. First, an aqueous slurry of starch (30-40% DS) is gelatinized (105 °C, 5 min) and partially hydrolysed (95 °C, 2 hours) by highly thermostable α-amylase to about DE 5-10. The optimum pH for the reaction is 6.0-6.5 and calcium (generally 50 ppm) is also needed. Then during saccharification, glucoamylase is added (60 °C, pH 4.0-4.5, 48 hours) to yield greater than 95-96% glucose, or with β-amylase (55 °C, pH 5.0-5.5, 72 hours) to yield around 80-85% maltose. Fleming [16] classified glucoamylase into two groups, out of which one converts starch and β-limit dextrins completely into glucose and the other one converts starch and β-limit dextrins into 80% and 40% glucose respectively. Recently Reilly [17] emphasized that the application of glucoamylase in starch industries may be benefited from thermostability and enhanced activity in the neutral pH range.

Glucoamylase alone is used to improve the efficiency of conversion of starch to glucose, or together with β-amylase to increase the maltose level in high maltose syrups [18]. Due to pH variation, large amounts of salts have to be removed by ion exchangers. Apart from the first step all other steps are time consuming, leading, in most cases, to reverse reactions and lower yields. Therefore, undesirable products like branched oligosaccharides, panose, isopanose and isomaltose are formed. The improvement in the starch conversion process is facilitated by finding new efficient and suitable enzymes possessing high thermostabilities, capable of working in acidic to neutral range of pH and independent of requirement of metal ion for activity. All these properties together can significantly lower the cost of sugar syrup production [1, 19].

The Ideal Process

Thermophilic organisms are important sources for the production of efficient thermostable enzymes. The production of enzymes that simultaneously attack α-1,4- and α-1,6-linkages may enhance the starch saccharification process. For replacement of traditionally used enzymes, it is of utmost importance to find more thermostable enzymes with unique properties [20]. Overproduction of such enzymes can be achieved by employing genetic techniques and optimization of fermentation processes [11, 21]. The enzymes with simultaneous α-1,4 and α-1,6 bond hydrolyzing capabilities are novel kind of pullulanases, and have been given various names such as amylase-pullulanase [22] and amylopullulanase [19].

Thermophiles and Thermostable Enzymes

Advantages of Thermostable Enzymes:

The thermophilic biocatalysts are important not only due to their thermostabilities but also because they are more resistant to denaturing agents and tolerant to higher solute (reactants) concentrations. A higher degree of thermophilicity of an organism does not necessarily imply that pure enzymes derived from such organisms will always be very thermostable. The fact remains, however, that one has a better chance of finding more thermostable or more chemoresistant enzymes in thermophiles than in mesophiles. High productivity is also expected from thermophiles. According to Arrhenius law, an increase in temperature speeds up chemical and enzymatic reactions, and therefore, microbial growth and product formation. Productivity is unfortunately given in terms of yield coefficients only, and not specific product formation rates, a parameter that allows easy and direct comparison between organisms [23].

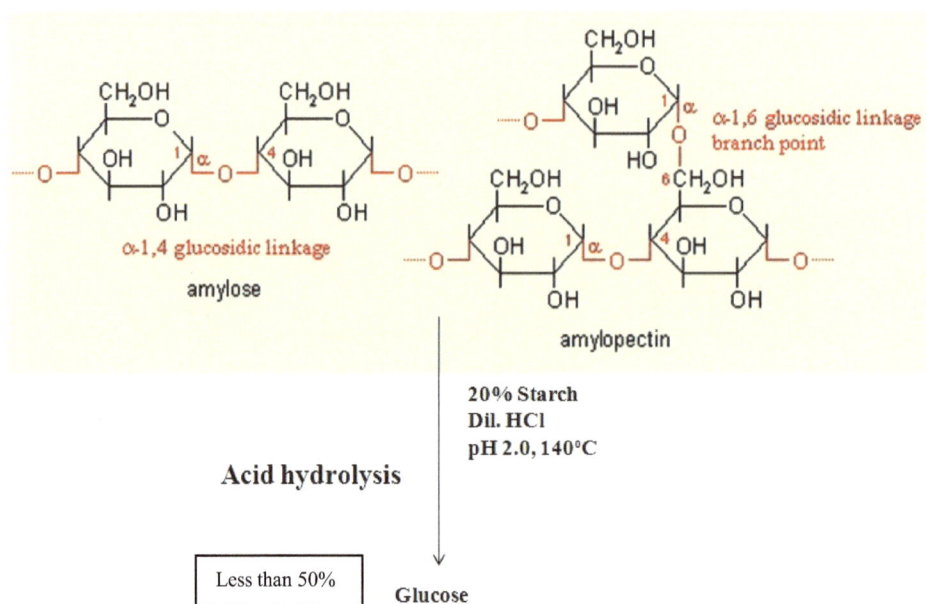

Figure 2: Acid hydrolysis of starch

The benefits of using enzymes as catalysts in industrial processes lie in their specificity and efficiency, leading to the production of by-products, less toxic wastes and reduced handling problems (Fig. **2**). The main disadvantages of using enzyme reside in stability problems and high cost. The latter partially depends on the former, since the frequency of replacement of enzyme in a bio-reactor (and therefore total production costs) is stability-dependent. Production cost itself is generally high due to low yields of enzyme per unit biomass, the expense of extraction and concentration and/or purification, losses of activity during purification and storage, and the handling-stability of the enzyme again being a factor [24, 25].

The use of thermostable enzymes reduces stability problems and in doing so alleviates some of the expense of production and replacement in a bioreactor. The stability of enzymes from thermophiles should lead to higher recoveries at ambient temperatures than is possible for mesophiles. The low activity of extremely thermophilic enzymes at ambient temperatures eases handling and storage problems, and the comparative molecular inflexibility that results in this inactivity at lower temperature has been suggested to lower the immune response to such proteins, thus reducing potential health risks [25]. Besides higher thermostability, the other expected advantages of thermophilic enzymes are increased chemo-resistance, a longer useful shelf life and less contamination problems [23] (Table **1**).

Table 1: Main Advantages of Thermostable Enzymes in Industrial Processes

S. No.	Property	Advantage in process
1	Thermostability	Tolerate high temp., last longer.
2	High optimum temperature	Little activity at low temperature, long shelf life.
3	Resistance to denaturing agents	Tolerate organic solvents, high and low pH
4	General robustness	Tolerate harsh purification, gives better yield.
5	Genes can be cloned in *E. coli*	A heating step makes purification easier.
6	Chemical reaction rates	Diffusion and other chemical processes are accelerated.
7	Solubility	Higher concentrations of poorly soluble compounds are possible.
8	Viscosity	Decreases; mixing & pumping can be accelerated; mass transfer rate increases.
9	Microbial contamination	Growth of all pathogens and most environmental mesophiles prevented.
10	Biological activity in raw materials	Heating kills most interfering enzymes or microbial activities.

Many of the enzymes presently used in industrial processes are quite thermostable even though they originate from mesophilic bacteria or fungi. Thermostability in enzymes from mesophiles is, however, the exception, not the rule. Since industrially useful enzymes must usually be thermostable, this characteristic is of primary importance in screening programs for enzymes. This is also important because the intrinsic basis underlying the thermostability of thermophilic enzymes is yet to be revealed and so engineering this characteristic into less thermophilic enzymes is not possible at this time. Successful cloning and expression of genes encoding hyperthermophilic enzymes in mesophilic hosts has improved the availability of high-temperature biocatalysts [26].

α-Amylases

The endo-acting α-amylases hydrolyse linkages in the interior of the starch polymer in a random fashion that leads to the formation of linear and branched oligosaccharides. Some of fungal α-amylases are listed in Table **2**. The sugar reducing groups are liberated in the α-anomeric configuration. Most starch hydrolyzing enzymes belong to the α-amylase family [27], which contain a characteristic catalytic $(\beta/\alpha)_8$-barrel domain. α-Amylase family is also known as family 13 glycosyl hydrolases [28].

This group includes the enzymes with the following features:

(i) They act on α-glycosidic bonds and hydrolyze this bond to produce α-anomeric mono-or oligosaccharides (hydrolysis), form α-1,4 or 1,6 glycosidic linkages (transglycosylation), or a combination of both activities.

(ii) They possess a $(\beta/\alpha)_8$ or TIM (Fig. **3**) barrel structure containing the catalytic site residues.

(iii) They have four highly conserved regions in their primary sequence, which contain the amino acids that form the catalytic site, as well as some amino acids that are essential for the stability of the conserved TIM barrel topology [29].

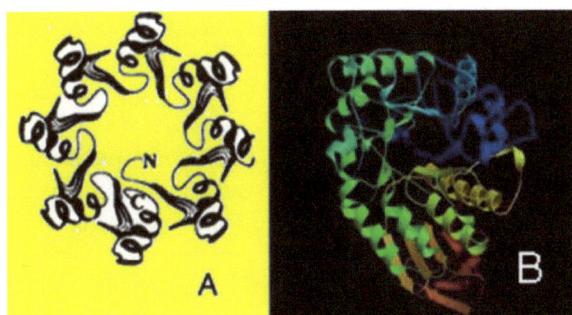

Figure 3: Schematic representation of the $(\beta/\alpha)_8$ barrel (A) and 3D structure of the α-amylase of *Aspergillus oryzae* or Taka amylase (B), obtained from protein database.

Starch saccharification is the synergistic action of a series of several amylases among them α-amylases of *Rhizomucor pusillus* and glucoamylase of *Thermomucor indicae-seudaticae* (E.C. 3.2.1.3) are the predominant bio-catalysts in starch saccharification [45]. α-Amylase hydrolyses starch, glycogen and related polysaccharides by randomly cleaving internal α-1,4-glucosidic linkages and liberates various lengths of oligosaccharides [1]. They are produced commercially in bulk from microorganisms (*Aspergillus* sp.) and represent about 25-33% of the world enzyme market, second place after proteases.

The applications of amylases are in the production of High Fructose Corn Syrup (HFCS), detergents, baking and ethanol [1, 11, 15, 46]. The global market for starch processing enzymes is around US $ 156 million and the cost of the enzymes used in the liquefaction process represented 24% of the total process cost [47]. Therefore, any improvement in the enzyme production/thermostability or activity will have a direct impact in the process performance, as well as in economics and feasibility [48]. Fungi have been widely used for the commercial production of thermostable α-amylases [49], an example is thermoacidophilic extracellular amylase from *Rhizomucor pusillus*. The most important characteristic of thermophilic organisms is their ability to produce thermostable enzymes with a higher operational stability and a longer shelf-life [50]. The α-amylases presently used in starch saccharification require Ca^{2+} for activity and /or stability. The compelling need for a novel α-amylase that does not require Ca^{2+} has been emphasized [1].

Table 2: Characteristics of Fungal α-Amylases

Name of the fungus	Temp Opt (°C)	pH Opt	K_m (mg ml^{-1})	Protein (Da)	Inhibitors	Reference
A. awamori ATCC 22342	55	5	1.0	54,000	Ag^{++}, Cu^{++}, Fe^{+++}, Hg^{++}	[30]
A. chevalieri NSPRI	60	5.5	0.19	68,000	EDTA, DNP	[31]
A. flavus LINK	60	6.0	500.0	52,500	Ag^{++}, Hg^{++}	[32]
A. foetidus ATCC 10254	45	5.0	2.19	41,500	-	[33]
A. fumigatus	55	6.0	-	-	-	[34]
A. hennebergi	50	5.5	-	50,000	-	[35]
A. niger ATCC 13469	50	5.0	-	-	-	[36]
A. oryzae	50	5.0	0.13%	-		[37]
A. oryzae M13	50	4.0	0.13%	52,000	-	
Aspergillus usamii	65	4.0	-	54,000	-	[38]
Cryptococcus S-2	60	6.0	-	66,000	-	[39]
Lipomyces kononenkoae CBS	70	5.0	800.0	76,000	DTT, Cu^{++}, Ag^{++}	[40]
Paecilomyces sp.	45	4.0	-	69,000	-	[41]
Thermomyces lanuginosus IISc91	55	5.6	2.5	42,000	-	[42]
Thermomyces lanuginosus	80	4.0	0.68	61,000	Zn^{++}	[43]
Trichoderma viride	60	5.5	-	-	-	[44]

FUNGAL α-AMYLASES

The α-amylases (alpha-1,4-glucan-4-glucanohydrolases, EC. 3.2.1.1) constitute a group of enzymes which catalyze hydrolysis of starch and other linear and branched 1,4-glucosidic oligo- and polysaccharides. A large section of worlds's population derives its food and energy either by directly utilizing starch or from its hyrolysates [51].

Sugar production by acid hydrolysis of starch was acceptable till 19 th century, is due to the scarce understanding of the proteinal benefits of biological catalysts. As the technology of enzymology progressed, a number of highly specific amylases are available to suit the commercial starch processing, the former having its own limitations including high temperature (40–150 °C) and low pH at 2.0 requirements, low glucose yields, formation of unwanted color, bitter tasting compounds, and the need for corrosion resistant vessels [15] was gradually replaced with biocatalysts. Now a days nearly 99.99% starch processing is based on enzymes.

The α-amylases are very thermostable and therefore suitable for processes carried out at high temperatures such as starch liquefaction in dextrose production processes. Another group of alpha-amylases are referred to as "Fungamyl®-like alpha-amylases", which are alpha-amylases related or homologous to the alpha-amylase derived from *Aspergillus oryzae*. The Fungamyl-like alpha-amylases have a relatively low thermostability, and the commercial product is sold under the tradename FUNGAMYL® by Novozymes A/S, Denmark. It has an optimum around 55 °C, and is not suitable for processes carried out at high temperatures. Fungamyl®-like alpha-amylases are today used for making syrups for, e.g., High Fructose syrup industry. Clearly, it would be advantageous to provide an alpha-amylase with increased thermostability preferably at an acidic pH. As far back as in 1980, Somkuti and Steinberg [49] described that a thermoacidophilic and extracellular alpha-amylase of *Rhizomucor pusillus* (*Mucor pusillus*), was isolated and characterized from soil. They state that "Since high temperature and acidic pH are optimum conditions for the economic hydrolysis of starch, the use of thermostable and acid-stable amylases of microbial origin for industrial purposes has been recommended", and they go on to conclude about the *Rhizomucor* amylase that: "It is apparently the first example of fungal alpha-amylase exhibiting both acidophily and thermophily simultaneously. Consequently, the alpha-amylase of *Rhizomucor pusillus* should be of economic importance [49].

Applications of α-Amylase

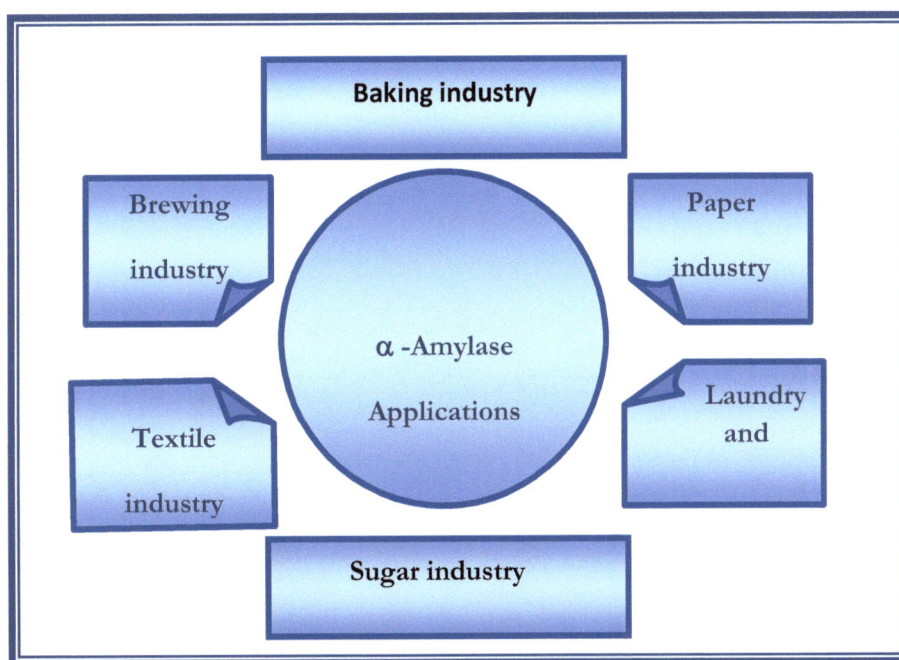

Figure 4: Various applications of α-amylases in different industries

Starch Liquefaction and Saccharification:

Starch is converted into high fructose corn syrups (HFCS) with the help of biocatalysts. Due to their high sweetening property, these are used in huge quantities in the beverage industry as sweeteners for soft drinks [47, 52] (Fig. **4** and **5**). The process requires the use of highly thermostable α-amylase for starch liquefaction and saccharification [53]. During liquefaction, starch granules are gelatinized in a jet cooker at 105 to 110 °C for 5 min in aqueous solution (pH 5.8 to 6.5) and then partially hydrolyzed using thermostable α-amylase at 95 °C for 2 to 3 hours. Incomplete starch gelatinization causes filtration problems in the downstream processing [54]. After liquefaction, the pH is adjusted to 4.2 to 5.0 and the temperature is lowered to 55 to 60 °C for the saccharification step. These pH adjustments increase chemical costs. They also create the need for ion-exchange refining of the final product to remove the added salts. So there is a need for neutral/acidic saccharifying α-amylases/ glucoamylases for the conventional saccharification step [54].

Figure 5: Some of the commercial sweeteners prepared by the amylases of fungi.

Table 3: The Comparative Analysis of Starch Hydrolysis by Acid and Enzymes

Typical Sugar Composition of Acid Converted Glucose Syrups			
Sugar	**30 DE**	**42 DE**	**55 DE**
Dextrose % of DS	10	19	31
Maltose % of DS	9	14	18
Trisaccarides % of DS	10	11	13
Higher sugars % of DS	71	56	48
Typical Sugar Composition of Acid Enzyme Converted Glucose Syrups			
Sugar	**28 DE**	**42 DE**	**63 DE**
Dextrose % of DS	5	6	37
Maltose % of DS	8	45	34
Trisaccarides % of DS	16	16	16
Higher sugars % of DS	71	33	13
Typical Sugar Composition of Enzyme Enzyme Converted Glucose Syrups			
Sugar	**High Maltose**	**Extra Maltose**	**High Liquid Dextrose**
Dextrose % of DS	3	3	95
Maltose % of DS	55	71	3

DE: Dextrose Equivalent
DS: Dextrose based solids

Hydrolysis carried out at high temperatures also minimizes polymerization of D-glucose to iso-maltose [55]. Amylolytic enzymes that produce specific malto-oligosaccharides in high yields from starch have received considerable attention. Such enzymes have a range of potential uses in the food (Fig. **5**), chemical and pharmaceutical industries [56].

The moderately thermostable saccharogenic α-amylase from *A. oryzae* is used industrially in the production of maltose syrups, and there is an increasing demand for high maltose syrups (Table **3**) [57]. Although maltogenic α-amylases that yield 53-80% maltose have been reported from Actionomycetes [58], their industrial potential is limited by their moderate thermostability and Ca^{2+} requirement. The use of Ca^{2+}-independent enzymes in starch hydrolysis eliminates the addition of Ca^{2+} in starch liquefaction and its subsequent removal by ion exchangers from the product streams [51, 55].

Detergent Industry:

Enzymes now comprise as one of the ingredients of modern compact detergents. The main advantage of enzyme application in detergents is due to much milder conditions than with enzyme free detergents. The early automatic dish washing detergents were very harsh, automatic dish washing detergents were not compatible with delicate china and wooden dishware. This forced the detergent industries to search for milder and more efficient solutions [59]. Enzymes also allow lowering of washing temperatures. α-Amylases have been used in powder laundry detergents since 1975. At present, 90% of all liquid detergents contain α-amylases [60]. One of the limitations of α-amylases in detergents is the enzyme stability in alkaline pH and low Ca^{2+} environment. Mostly amylases are unstable with detergents and oxidative agents. Recently scientists from the two major detergent enzyme suppliers, Novozymes and Genencor International have used protein engineering to improve the detergent and bleach stability of the amylases [61, 62]. Genencore International and Novozyme have introduced these new products in the market under the trade names Purafect OxAm® and Duramyl®, respectively.

Paper Industry:

The use of α-amylase for the production of low viscosity, high molecular weight starch for coating of paper is reported [63]. As for textiles, sizing of paper is performed to protect the paper against mechanical damage during processing. It also improves the quality of the finished paper. The size enhances the stiffness and strength in paper. It also improves the erasability and also provides good coating for the paper. Starch is also a good sizing agent for the finishing of paper. Starch is added to the paper in the size press and paper gets impregnated by the starch by passing through two rollers that transfer the starch slurry. The temperature of this process lies in the range of 50-65 °C. The viscosity of the natural starch is too high for paper sizing, and is adjusted by partially degrading the polymer with thermostable α-amylases.

A number of amylases exist for use in the paper industry, which include Amizyme® (PMP Fermentation Products, Peroria, USA), Termamyl®, Fungamyl, BAN® (Novozymes, Denmark) and α-amylase G 9995® (Enzyme Biosystems, USA).

Baking Industry:

For, decades, enzymes such as malt and microbial α-amylases have been widely used in the baking industry [64, 65]. These enzymes were used in bread and allied products to give high quality products having better color and softer crumb. Many enzymes such as proteases, lipases, xylanases, pullulanases, pentosanases, cellulases, glucose oxidase, lipoxygenase etc. are being used in the bread industry for varied purposes, but none had been able to replace α-amylases.

Up to now, the α-amylases used in baking have been cereal enzymes from barely malt and microbial enzymes from fungi [66, 67]. α-Amylase supplementation in flour not only enhances the rate of fermentation and reduces the viscosity of dough (resulting in improvements in the volume and texture of the product), but also generates additional sugar in the dough, which improves the taste, crust color and toasting qualities of the bread [68].

Upon storage, the crumb becomes dry and firm, the crust loses its crispness, and consequently the flavor of the bread deteriorates. All these undesirable changes in the bread are together known as staling. The improvement of retrogradation of starch fraction in bread staling has been emphasized [69]. A loss of more than US$1 billion is incurred in USA alone every year due to the staling of bread.

Conventionally various additives are used to prevent staling and improve the texture and flavor of baked products. Additives include chemicals, small sugars, enzymes with/ without other combination, milk powder, emulsifiers, monoglycerides or diglycerides, sugar esters and lecithin [70]. Recently emphasis has been laid on the use of enzymes in dough improvement/as anti-staling agents, e.g. α-amylase [71, 72]. Pullulanases and α-amylase combination are used for efficient antistaling [73].

Bioethanol Production:

For thousands of years, ethanol has been produced for human consumption, and for at least a thousand years it has been possible to make concentrated alcoholic drinks by means of distillation. Ethanol for use as a chemical feedstock was produced by fermentation in the early days of industrial microbiology; however, for many instead, primarily through the catalytic hydration of ethylene. In recent years, attention has turned again to the production of ethanol for chemical and fuel purposes by fermentation process.

In Brazil, as early as 1982, about 30% of petroleum imports were replaced by ethanol production from sugar cane, involving 60-80 separate fermentation plants. In 1986, in the United States, 65 separate fermentation plants were used in ethanol operation. Bioethanol, which is produced from biomass resources by fermentation, is the most promising biofuel and the starting material of various chemicals. In the US, ethanol produced from corn starch has already been used as biofuel and production volume has increased rapidly. However, processes to reduce the high production costs are required. The production of fuel ethanol by fermentation requires the ability to produce high ethanol concentrations rapidly while maintaining good yields. Rapid fermentation and high alcohol levels are desirable to minimize capital costs and energy required for distillation, while good yields are necessary for economic viability.

Traditionally, ethanol fermentation relies on sugar-rich substrates, mainly sugarcane, because their carbohydrate is in fermentable form. However, sugarcane is an expensive material and not continuously available because it is a seasonal crop [74]. Thus there are great economic advantages in extending the substrate range of ethanol-fermenting microorganisms so that the ethanol may be produced from cheap substrates such as starchy crops [75] and cellulosic materials [75-77].

Ethanol-fermenting microorganisms, such as *Saccharomyces cerevisiae* and *Zymomonas mobilis* lack amylolytic enzymes and are unable to directly convert starch into ethanol. Traditionally, starch is hydrolysed enzymatically into fermentable sugar *via* liquefaction and saccharification processes prior to ethanol fermentation [45].

In the process currently employed for industrial-scale ethanol production from starchy materials, starch is first hydrolysed by adding a liquefying enzyme namely α-amylase, and then cooked at high temperature. The liquefied starch is then hydrolysed to glucose with a saccharifying enzyme glucoamylase, and glucose is fermented to ethanol by yeast cells.

Miscellaneous Applications:

α-Amylase, pullulanase, cyclodextrin glucosyltransferase, and maltogenic amylase are nowadays widely used by industry in various applications. Among them α-amylase probably has the most widespread use. Besides their use in the saccharification or liquefaction of starch, these enzymes are also used for the preparation of viscous, stable starch solutions used for the warp sizing of textile fibers, the clarification of haze formed in beer or fruit juices, or for the pretreatment of animal feed to improve the digestibility. A growing new area of application of α-amylases is in the fields of laundry; textile desizing and dish-washing detergents. A modern trend among consumers is to use lower temperatures for doing the laundry or dishwashing. The removal of starch from porcelain becomes more problematic. Detergents with α-amylases optimally working at moderate temperatures and alkaline pH can help to solve this problem [51].

FUNGAL GLUCOAMYLASES

The applications of glucoamylases (E.C. 3.2.1.3, glucan 1,4 -α-glucosidase) in starch saccharification lies in sugar industry is due to its ability to release glucose as the major end product, which is used in food, beverage, ethanol, amino acids, organic acids [78]. Glucoamylase is one of the enzymes of world wide interest in starch saccharification to yield glucose for use in food and fermentation industries. Glucoamylase is one of the high demand commercial biocatalysts in food industry, which is required in higher tonnage than almost any other enzyme [17, 79].

Industrially glucoamylases are produced from filamentous fungi, *Aspergillus* and *Rhizopus* spp. *via* submerged fermentation using production medium with a typical concentration of 20% corn and 2.5% of corn steep liquor at 30 to 35 °C [56]. The fungal glucoamylases being optimally active at around pH 4.0 to 4.5, the saccharification is essentially carried out under acidic conditions at 60 °C for 3-4 days to achive a final yield of 98% glucose [47, 80]. Glucoamylase is one of the high demand commercial biocatalysts in food industry, which is required in higher tonnage than almost any other enzyme [17, 79]. Industrially glucoamylase is produced from filamentous fungi, *Aspergillus* and *Rhizopus* spp., *via* submerged fermentation using production medium with a typical concentration of 20% corn and 2.5% corn steep liquor at 30-35 °C [56, 81]. A number of glucoamylase production media for a variety of microbes have been optimized at laboratory scale including liquid, solid as well as aqueous two-phase systems [12, 45, 82, 83]. Different fermentation vessels, from shake flasks to bioreactors, and strategies including batch, fed-batch and continuous fermentations have been employed [45, 84, 85, 86]. Glucoamylase production by *Thermomyces lanuginosus* was found to be 2.5-fold higher in shake flasks compared to static cultures [87]. Some of the fungal glucoamylases characteristics are shown in Table **4**.

Table 4: Characteristics of Fungal Glucoamylases

Name of Fungi	Post-translational modification	Molecular weight (kDa)	Purification procedure	pH	Temp (°C)	Km starch	Ref.
Aspergilus niger Bo-1, *Monascus* sp.	(-)	91, 73, 59	Ultra-filtration and ion exchange			-	[88]
Humicola sp.	(-)	72.8	Anion exchange hydrophobic interaction chromatography	4.7	55	0.26 mg/mL	[89]
Fusarium solani	(-)	40, 41	Ammonium sulfate precipitation, gel filtration, anion exchange, and hydrophobic interacton chromatography	4.5	40	1.9 mg/mL	[90]
Chaetomium thermophilum	Glycoprotein (11.7%)	64	Ammonium sulfate precipitation, anion exchange, hydrophobic interaction chromatography	4.0	65	-	[91]
Scytalidium thermophilum	Glycoprotein (9.8%)	75	Ion exchange chromatography	5.5	70	0.21 mg/mL	[92]
Scytalidium thermophilum	Glycoprotein (25.5%)	86	Ion exchange chromatography	6.5	60	0.28 mg/mL	[93]
Thermomyces lanuginosus	Glycoprotein (3.27%)	75		4.4-5.6	70	0.80 mg/mL	[94]
Curvularia lunata	(-)	66	Ammonium sulfate precipitation, gel filtration, and ion exchange	4.0	50	-	[95]
Therermomyces lanuginosus	(-)	66		5.0	70	3.5 mM	[96]
A. oryzae	Glycoprotein (7.8%)	68.4	Ethanol precipitation and affinity chromatography using acarbose	4.0-4.5	56	-	[97]
Thermomucor indicae-seudaticae	Glycoprotein (9-10.5%)	42	Ion exchange and gel filtration	7.0	60	0.45 mg/mL	[98]
Arthrobotrys amerospora ATCC 34468	(-)	44.7, 71, 74.5	Ion exchange and gel filtration	5.6	55	-	[99]
Paecilomyces variotii	Glycoprotein (27.5%)	86.5	Ion exchange and gel filtration	5.0	55	3.8-4.1 mg/mL	[100]

Applications of Glucoamylases

The major commercial application of glucoamylase is to catalyze starch, to yield glucose for use in food and fermentation industries. Glucose production from starch, along with glucoamylase requires the synergistic action of a series of amylases. Some of the commercially important fungal glucoamylases are given in the Table **5**. In the First step, ~30-35% dry solid starch slurry is gelatinized (~60-90 °C) and subsequently liquefied at 95-105 °C (pH 6.5) by alpha-amylase to short-chain dextrins. These dextrins in the next step are saccharified by glucoamylase to release glucose. The fungal glucoamylase being optimally active at around pH 4.0-4.5, the saccharification is essentially carried out under acidic conditions at 60 °C for 3-4 days to achieve a final yield of ~96% glucose [17, 47].

Additionally, debranching enzymes (pullulanase or isoamylase) are used to hasten starch processing by cleaving alpha-1,6 glycosidic bonds, which allows to attain an early peak in glucose yield with less byproduct formation [80]. Glucose has ~75% of the sweetness of sucrose, while its isomer fructose is twice sweeter than sucrose. Consequently, fructose is preferred especially in so-called low calorie health/diet foods, where it provides double the sweetness of sucrose at half the weight and can be metabolized without insulin [101]. Commercially fructose is produced by the isomerization of glucose using fungal glucose/xylose isomerase (E.C. 5.3.1.5, D-xylose-ketol isomerase) at 50-60 °C and pH 7-8. Glucose isomerase is the most expensive of all the enzymes involved in starch processing, and thus, is reused until it loses most of its activity. The concentrated glucose syrup is passed through the immobilized glucose-isomerase column or sometime through the glucose isomerase producing cells [47]. The process yields around 40-42% of fructose and the concentration of fructose in the final product is raised to ~55% by chromatographic enrichment of glucose-fructose mixture [17, 102].

Fungi in Ethanol Production:

In the present day biotechnology for ethanol fermentation, especially Japanese are using rice as a raw substrate. They inoculated *Aspergillus oryzae* to grow on the upper suface of the rice. The fungus converts the starch into simpler oligosaccharides, which can be utilized by yeast to produce ethanol.

Table 5: The Commercially Important Fungal Glucoamylases

Name of fungus	Name of the compound	Industry	Contact address
Aspergillus niger mutant	-	China-America Technology Corp.	Marshak Science Bldg., Rm. J423, 138th St. at Convent Ave., New York, NY 10031
-	Glucoamylase liquid 25KG /DRUM	Jinzhu Tibet Co., Ltd.	6F Jinzhu Mansion, 10# JiuXing Ave. Hi-Tech Dist. Chengdu. Sichuan, China 610041
Aspergillus niger	ENZECO® GLUCO AMYLASE-L 10001	Enzyme Development Corporation	21 Penn Plaza, 360 West 31st Street, New York, NY Street, Chengdu, Sichuan, China 610031
	Liquid/Solid Glucoamylase	Sichuan Shan Ye BioTech Co., Ltd.	No.2-12-14, Platinum Building, No.27, Qinglong, China
-	Syder Brand Glucoamylase	Wuxi Syder Bioproducts Co., Ltd.	No 30 Huaxia Road, Xishan District Wuxi, Jiangsu China 214000
Aspergillus niger	Liquid/Solid Glucoamylase	Shanghai Kaiquan Biotechnology Co., Ltd.	Rm 2505,Minzhu Building, 1328 Zhangyang Road, Pudong, Shanghai, China 200122
Aspergillus niger	Glucoamylase Sunson GA-L, GA	Sunson Industry Group Co., Ltd.	Suite 628, China Minmetals Building, Block 4 Anhuili, Chaoyang District, Beijing, China
Aspergillus niger	KDN-GE01TM	Qingdao Continent Industry Co., Ltd.	Room 0607, Aucma Mansion, 29 Miaoling Road, Qingdao City, Shandong Province. China
-	DEXTRO 300L	Advanced Enzyme Technologies Ltd.	A' Wing, Sun Magnetica, 5th Floor, Accolade Galaxy, LIC Service Road, Louisewadi, "ane-400 604, India
-	Palkodex	Maps (India) Limited	302, Shapath-3, Nr. GNFC Info Tower, S.G. Road, Ahmedabad-380 054, India
-	Dextrozyme® DX/GA/W	Novozymes	Novozymes Krogshoejvej 36, DK-2880 Bagsvaerd
Aspergillus niger	STARGEN™ enzyme blend (alpha-amylase/glucoamylase)	Genencor International	Danisco A/S Langebrogade 1 DK-1001 Copenhagen
Aspergillus niger, *Rhizopus niveus*, and *R. delemar*	Gluczyme AF6	Amano Enzymes USA Co., Ltd.	2150 Point Boulevard, Elgin, IL 60123

Murai *et al.* [103] reported that the fast growth of *S. cerevisiae*, was co-cultivated with *Rhizopus oryzae* of glucoamylase and *Bacillus stearothermophilus* of α-amylase on surface of the starch, which yielded good production of sugars. In another report, the recombinant *S. cerevisiae* YPB-G expressing the *Bacillus subtilis* α-amylase and *A. awamori* glucoamylase was shown to produce higher amounts of ethanol in the production medium containing 40 g/L of starch [104].

PECTINASES FROM FUNGI

Pectinases were some of the first enzymes to be used in homes. Their commercial application was first observed in 1930 for the preparation of wines and fruit juices. Only in the 1960s did the chemical nature of plant tissues become apparent and with this knowledge, scientists began to use a greater range of enzymes more efficiently. As a result, pectinases are today one of the upcoming enzymes of the commercial sector. Primarily, these enzymes are responsible for the degradation of the long and complex molecules called pectin that occur as structural polysaccharides in the middle lamella and the primary call walls of young plant cells. Pectinases are now an integral part of fruit juice and textile industries as well as having various biotechnological applications.

Structure of Pectin

Chemically, pectic substances are complex colloidal acid polysaccharides, with a backbone of galacturonic acid residues linked by α (1–4) linkages. The side chains of the pectin molecule consist of L-rhamnose, arabinose, galactose and xylose. The carboxyl groups of galacturonic acid are partially esterified by methyl groups and partially or completely neutralized by sodium, potassium or ammonium ions. Based on the type of modifications of the backbone chain, pectic substances are classified into protopectin, pectic acid, pectinic acid and pectin [105].

Classification of Pectic Enzymes

Pectinases are classified under three headings according to the following criteria: whether pectin, pectic acid or oligo-D-galacturonate is the preferred substrate, whether pectinases act by trans-elimination or hydrolysis and whether the cleavage is random (endo-, liquefying of depolymerizing enzymes) or endwise (exo- or saccharifying enzymes). The three major types of pectinases are:

Pectinesterases (PE): Pectinesterases also known as pectinmethyl hydrolase, catalyzes deesterification of the methoxyl group of pectin forming pectic acid. The enzyme acts preferentially on a methyl ester group of galacturonate unit next to a non-esterified galacturonate unit.

Hydrolyzing glycosidic linkages: Polymethylgalacturonases (PMG): catalyze the hydrolytic cleavage of α-1,4-glycosidic bonds.

1. *Endo-PMG*: causes random cleavage of α-1,4-glycosidic linkages of pectin, preferentially highly esterified pectin.

2. *Exo-PMG*: causes sequential cleavage of α-1,4-glycosidic linkage of pectin from the non-reducing end of the pectin chain.

Polygalacturonases (PG): catalyze hydrolysis of α-1,4-glycosidic linkages in pectic acid (polygalacturonic acid). They are also of two types:

1. *End-PG*: also known as poly (1,4-α-D-galacturonide) glycanohydrolase, catalyzes random hydrolysis of α-1,4-glycosidic linkages in pectic acid.

2. *Exo-PG*: also known as poly(1,4-α-D-galacturonide) galacturonohydrolase, catalyzes hydrolysis in a sequential fashion of α-1,4-glycosidic linkages on pectic acid.

Cleaving α-1,4-glycosidic linkages by trans-elimination, results in galacturonide with an unsaturated bond between C4 and C5 at the non-reducing end of the galacturonic acid formed.

Polymethylegalacturonate lyases (PMGL): they breakdown of pectin by trans-eliminative cleavage

1. *Endo-PMGL:* also known as poly(methoxygalacturonide) lyase, catalyzes random cleavage of α-1,4-glycosidic linkages in pectin.

2.   *Exo-PMGL:* catalyzes stepwise breakdown of pectin by trans-eliminative cleavage.

Polygalacturonate lyases (PGL)

Catalyze cleavage of α-1,4-glycosidic linkage in pectic acid by trans-elimination. They are also of two types:

1.   *Endo-PGL*: also known as poly (1,4-α-D-galacturonide) lyase, catalyzes random cleavage of α-1,4-glycosidic linkages in pectic acid.

2.   *Exo-PGL*: also known as poly (1,4-α-D-galacturonide) exolyase, catalyzes sequential cleavage of α-1,4-glycosidic linkages in pectic acid.

Applications of Fungal Pectinases

Pectinases in Fruit Juice Extraction:

They have widespread applications in the textile, food, plant tissue maceration, wastewater treatment and degumming of natural fibers [106, 107]. Pectinases, with combination of hemicellulase and cellulase are used for cell wall destruction and fruit juice extraction [108]. Where as, the pectin esterase, PGase and pectin lyase have been using in fruit juice depectinization, clarification and extraction. The production of pectinolytic enzymes has been widely explored in filamentous fungi [109]. However, there are a very few reports on the production of pectinases by thermophilic moulds [110]. Pectinases hydrolyse insoluble pectin that imparts gelling power in the apple pomace during pressing. Moreover, pectin is the primary cause of juice viscosity and maintains the particles in suspension making it difficult for electrostatic charges to develop between pectin, proteins and tannins that must occur to facilitate clarification [111, 112]. Pectinase preparations containing numerous side activities are added to the pulp in the maceration step during which cell walls are partially hydrolysed reducing the viscosity drastically. Pectinases are being used to reduce viscosity, consistency and turbidity, thereby increasing juice yield and allowing easier clarification and concentration, and increasing the pressing speed and liberation of reducing sugars [113].

The hemicellulases in combination with pectinases cause only a minimal increase in the release of sugars, and these are important in the liquefaction process, since these facilitate the separation of juice from the solids. Soluble hemicelluloses like arabans can also give haziness to concentrates. After crushing fruits, the crude juices are often very viscous and the remaining solids are difficult to separate from the juice [114]. The Grassin and Fauquembergue [113] reported that the presence of cellulase in the mixture of enzymes exerted a favorable effect in improving the viscosity and filterability of banana juice [109]. The effects of Pectinase on quality of carrot juice were also studied. It was found that pectinase treatment can not only improve the juice yield but also enhance total carotene content in the juice without harming the cloudy stability of juice [115]. The exogenous enzymes reinforce the low levels of the activities of endogenous enzymes present in fruits. A new technology of total liquefaction may be developed by using pectinases, hemicellulases and cellulases in combination for obtaining a clarified and depectinized juice in a single step by Kaur *et al.* [109]. The addition of exogenous enzymes also allows more specific degradation that is necessary to give smooth texture to juice, not found with heating, and at the same time preserves color and vitamins [113]. There was an increase, compared to the control, in the volume of juice of banana, grapes and apple caused by the treatment of fruit pulps with the enzymes of *S. thermophile* [109].

Coffee and Tea Fermentation:

Pectinases play an important role in coffee and tea fermentation, especially to remove the mucilage coat from the coffee beans and also to remove the pulpy layer of the bean. Cellulases and hemicellulases present in the enzyme preparation aid the digestion of the mucilage. A diluted commercial enzyme preparation is sprayed on to the beans at a dose of 2–10 g per ton at 15–20 °C. The fermentation stage of coffee processing is accelerated and reduced from 40 to 80 hours to about 20 hours by enzyme treatment. Since large-scale treatment of coffee with commercial pectinases is costly and uneconomical, inoculated waste mucilage is used as a source of microbial pectic enzymes. The fermentation liquid is washed, filtered and then sprayed on to the beans [116].

Fungal pectinases are also used in the manufacture of tea. Enzyme treatment accelerates tea fermentation, although the enzyme dose must be adjusted carefully to avoid damage to the tea leaf. The addition of pectinase also improves the foam-forming property of instant tea powders by destroying tea pectins [117]. The important producers of these pectinases as reported in the literature are given in the Table **6**.

Table 6: Characteristics of Fungal Pectinases

Name of fungus	Mode of action	Optimum pH	Optimum temperature (°C)	References
Aspergillus niger	Endo-pectinase, Exo-pectinase	4.5-6.0 3.5-5.0	Below 50	[118]
Mucor pusillus	PG	5.0	40	[119]
Penicillium frequentans	Endopolygalacturonase (Endo-PG)	4.5-4.7	50	[120]
Rhizoctonia solani	Endo-PG	4.8	50	[121]
Sclerotium rolfsii	Endo-PG	3.5	55	[122]
Sporotricum thermophile	Endo-PG and Exo-PG	5.0-7.0	45	[109]

FUNGAL PHYTASES

Phytases (myo-inositol hexakisphosphate phosphohydrolase; EC 3.1.3.8 and EC 3.1.3.26) belong to the family of histidine acid phosphatases (HAP), a subclass of phosphatases, which catalyze the hydrolysis of phytic acid, the principal storage form of phosphorus in cereals, legumes, oil seeds, nuts and others. The fungal cultures are employed for the production of phytases in SmF or SSF [123]. Maize starch-based medium was used for the production of phytase in SmF by using *Aspergillus niger*. Activity of the enzyme was found to be 1.075 phytase units per min per ml of the crude culture filtrate at pH 5.5 and 40 °C (10 days). Extra-cellular phytase produced by *A. niger* 5990 showed a fivefold higher activity in liquid culture when compared with cultures of *A. ficuum* NRRL 3135. SmF was carried out at 35 °C, pH 7 for four days. The phytase had a higher optimum temperature for its activity than the commercial enzyme, Natuphos, from *A. ficuum* NRRL 3135 [124]. SSF was employed for phytase production using strains of *Aspergillus* sp. [125] used canola meal for phytase production by *A. ficuum*. Optimum substrate moisture was 64%. Age of the inoculum had a profound effect on enzyme synthesis by the culture. For cultivation of *A. carbonarius* on canola meal was founded 53–60% moisture [126]. However, addition of glucose at lowers concentrations (6 g) and surfactants such as Na-oleate or Tween-80 in the medium increased biomass growth and enzyme synthesis [127, 128].

Applications of Phytases

Phytases in Fish Feed:

Phytate is the main storage form of phosphorus (P) in many plants, but phytate-bound P is not available to monogastric or agastric fish animals [129]. Phytase, an enzyme specific to hydrolyze indigestible phytate, has been increasingly used in fish feed during the past two decades [130-132], mainly in response to heightened concerns over P pollution to the aquatic environment. Since global phosphate reserves are not renewable, phytate-P as an alternative and economical P source can be effectively converted to available-P by phytase. The capability of this enzyme to enhance bioavailability of P and reduce P load is well documented [133]. Phytase supplementation also leads to improved availability of other minerals and trace elements [134-136]. Nevertheless, there is still no consistent conclusion that phytase could enhance protein and energy utilization. Studies in amino acid digestibility after phytase supplement are mutative and the underlying mechanisms have not been fully understood. Because phytase is very sensitive to pH and temperature, the utilization of phytase in fish feed is still on its first stage compared with that of in poultry and swine feed. A wide variety of phytases were discovered and characterized in order to find the optimum enzyme which is stable in application, resistant against high temperatures, dust-free, and easy to handle. Initial steps to produce phytase in transgenic plants and fish animals are also undertaken. In this review [133], the authors focus on comparing properties of phytase from different sources, examining the effects of phytase on P utilization and aquatic environment pollution, meanwhile providing commercial potentiality and impact factors of phytase utilization in fish feed [133].

Other Applications:

Phytase has been used as a cereal feed additive to enhance the phosphorus and mineral absorption in monogastric animals and to reduce the anti-nutritional properties of phytic acid. Phytases catalyze hydrolysis of phytic acid in a stepwise manner to inositol phosphates, myo-inositol and inorganic phosphate utilizing a phosphohistidine intermediate in their phosphoryl transfer reaction [137]. In grains, nearly 50–80% of phosphorous is tied up in phytins (Ca^{2+}/Mg^{2+} salts of *myo*-inositol hexakisdihydrogen phosphate), which are not digested by monogastrics due to lack of adequate levels of phytase in their digestive tracts. The phytins thus wind up in manure and liquid effluent, where it is degraded by microorganisms leading to environmental pollution [138-140]. The role of phytin–phosphorus in the plant was earlier speculated as a storage product. It is believed that a large amount of phosphorus is stored in the seed and it gets liberated on germination and incorporated into ATP. Recent studies have established the role of inositol phosphate intermediates in the transport of materials into the cell and as secondary messengers in signal transduction [141]. Due to the interaction of phytic acid with other compounds, it acts as an antinutritional factor in several ways: (i) six reactive groups in the molecules of IP6 make it a strong chelating agent, which binds cations such as $Ca^{2+,}$ Mg^{2+}, Fe^{2+}, Zn^{2+}. Under gastrointestinal pH conditions, insoluble metal phytate complexes are formed which make the metals unavailable for absorption in the intestinal tract of animals and humans [142], (ii) phytates reduce digestibility of proteins, starch and lipids by forming insoluble complexes, and (iii) the action of certain enzymes such as amylase, trypsin, acid phosphatase and tyrosinase is inhibited by phytic acid and inositol pentaphosphate [143]. There are several methods for the reduction of phytic acid such as cooking, autoclaving, ion exchange, elution with buffer and germination of seeds. The use of these methods, in most cases, leads to loss of nutritional constituents. An alternative is the enzymatic treatment method using phytases is considered to be superior to the others. Ruminant animals sustain the microflora that enzymatically release inorganic phosphorus from phytic acid. Monogastric animals such as humans, chickens and pigs however produce little or no phytase in the intestine. Hence, the phytic acid phosphorus is unavailable and the phytin-P is excreted [144, 145]. Phytic acid present in the manure of these animals is enzymatically cleaved by soil and water-borne microorganisms and the phosphorus thus released is transported into the water bodies causing eutrophication. This results in oxygen depletion due to excessive algal growth, which lowers the dissolved oxygen in water bodies resulting in fish kills.

Supplementation of animal feeds with phytase provides swine and poultry producers with a safe and effective management tool to reduce nutrient run off by significantly reducing the amount of phosphorus excreted in the manure of the animals [146-148]. Furthermore, the reduction or elimination of inorganic phosphorus supplementation of animal feed reduces P in the manure by more than 33%, thus cutting the pollution burden by one-third [149]. This problem can be overcome by adding phytases (*myo*-inositol hexakisphosphate 3-phosphorylase, EC 3.1.3.8 and *myo*-inositol hexakisphosphate 6-phosphorylase, EC 3.1.3.26) to monogastric animal feed that improves the nutritional status of foods and help in combating environmental phosphorus pollution [150]. The main problem with the widespread application of phytase as a feed supplement is the current high price of commercial phytase, which may add US$2–3 per metric ton to the feed cost [151]. The enzyme is currently produced by conventional submerged fermentation (SmF), a more expensive process [152]. The phytase production by *S. thermophile* has reported in submerged and solid state fermentations and the enzyme reduced efficiently the phytate content in sesame oil cake, wheat flour, soymilk and during bread making [153-157].

FUTURE PERSPECTIVES

Despite the fact that α-amylases have been used in a wide variety of technical applications for several years, there were few new developments. The available enzymes are good and useful for the needs of the customers. The interest in new and improved α-amylase is growing, and consequently the research is intensified as well. From the description of the High Fructose Corn Syrups (HFCS) process it can be concluded that there is a need for an α-amylase that has the same thermostability and also independent of Ca^{2+}. Such an enzyme would fit much better into the process, because the addition and removal of the metal ions before and after the liquefaction step could be avoided. The natural pH of starch slurry is approximately 4.5. The present starch-processing methods require adjusting the pH to 5.8 or above and then reducing it to 4.2 to 4.5 for the saccharification step. These two pH adjustments increase chemical costs. They also create the need for ion-exchange refining of the final product to remove the added salts. An α-amylase able to work at lower pH would reduce these costs, simplify the process, and reduce the formation of high-pH by products (while produced in the enzyme catalyzed reaction mixture) (e.g.

maltose) [47]. Further, there is a need for Ca^{2+}-independent and thermostable α-amylases, which are functional around 100 °C at low pH values for improved processing procedures. Increasing the saccharification process temperature would result in many benefits such as higher substrate concentrations, decreased viscosity and lower pumping costs, limited risks of contaminations, increased reaction rates and decrease of operation time, lower costs of enzyme purification and longer catalyst half-life, due to increased enzyme thermostability. On the other hand, for the detergent industry a better compatibility with the detergent formulations is necessary. A detergent enzyme has to resist the oxidative potency of the bleach system in detergents, the low calcium concentration and high concentration of surfactants in the wash liquor. The stability against oxidative agents in detergents has been addressed by two companies and the stability at low Ca^{2+} ion concentrations by another. It can be expected that the protein engineering efforts for the improvement of this enzyme will increase, since the recent determination of its X-ray structure will allow rational protein engineering [158].

The first protein engineering attempts were successful even without an X-ray 3D-structure. The problem of the calcium dependence was solved by using a hypothetical 3D-model, which was obtained by homology modeling using the X-ray structure of the α-amylase from *A. oryzae* [159] as a template [160]. Based on this model, the region around the structural calcium-binding site was investigated for positions where a negative charge could be introduced or a positive charge could be removed by amino acid substitutions. The rationale behind this approach was to increase the negative potential around the calcium-binding site to make this area more attractive for positive ions like calcium. As a result the calcium ions should be bound more tightly in the binding site giving a more thermostable enzyme. The storage stability of the variant enzyme in bleach containing detergents improved as compared to the wild-type enzyme. The performance of the mutated enzyme in bleach containing detergent formulations is much better than the performance of the wild-type enzyme. The relation between stability of enzymes against detergent ingredients and their performance has been reported [161].

Directed evolution has rapidly emerged as a powerful new strategy for improving the characteristics of enzymes in a targeted manner. By coupling various protocols for generating large variant libraries of genes, together with high-throughput screens that select for specific properties of an enzyme, such as thermostability, catalytic activity and substrate specificity, it is now possible to optimize biocatalysts for specific applications [162].

During the last three decades, α-amylases have been exploited by the starch processing industry as a replacement for acid hydrolysis in the production of starch hydrolysates. This enzyme is also used for removal of starch in beer, fruit juices, or from clothes and porcelain. A new and recent application of maltogenic amylase is as an anti-staling agent to prevent the retrogradation of starch in bakery products. Out of the vast pool of enzymes, glucoamylases are one of the oldest and world widely used enzymes in food industry. The major applications of glucoamylases are in starch saccharifaction for production of glucose, which is an essential substrate for various fermentation processes and beverage industries. Traditionally glucoamylases have been produced from filamentous fungi, is due to they secrete large quantities of the enzyme extracellularly. The commercially used fungal glucoamylases have certain limitations such as moderate thermostability, acidic pH requirement, and slow catalytic activity that increase the process cost. Consequently, the search for newer glucoamylases and protein engineering to improve pH and temperature optima leading to amelioration in catalytic efficiency of existing enzymes have been the major areas of research over the years [78]. The commercial demands are so high that even small improvements in production and catalytic efficiency could prove lucrative. The modern mutagenesis and genetic engineering techniques together have largely solved the production problems, with improved strains being able to secrete as high levels 40 g per liter of the production medium.

Fungal metabolites are being used in the spheres of food, feed and therapeutics. Since most of these saprotrophic and mycorrhizal organisms can be genetically modified with ease, research initiatives like EUROFUNG currently underway have taken up strain improvement programmes with state of the art technologies to accomplish a unified objective of developing "secretion giants" out of these modest high potential organisms. In this regard, the optimization of production process deserves parallel importance. Although most current methods rely on submerged fermentation (SmF), the use of traditional solid state fermentation (SSF) processes should be explored more thoroughly. Several results have indicated that the SSF process results in improved levels of various secreted fungal hydrolases. Until now, most of the research in the field of SSF has been focused on process and fermentor design [163], treating the organism involved as a black box so there is need for the development for this black box to be fully organised.

At the ending of the 20th century, annual sales of fungal phytase (an animal feed and in pond water ecosystem) as additive were estimated to be $500 million. Evolution of the market for this feed additive can be attributed to a chain of events in the end of 20th century that created the demand for the enzyme, and thus, provided a means for its commercial development to be organised in good manner [164].

In the light of modern biotechnology, fungal enzymes are now gaining importance in biopharmaceutical applications. Their application in food and feed based industries is the major market, and further the demand for α-amylases, glucoamylases, pectianses, and phytases would always be increasing in these biotechnology sectors.

REFERENCES

[1] Antranikian, G. Microbial degradation of starch In: Winkelmann G, Ed. Microbial Degradation of Natural products, Weinheim VCH: Germany. 1992; pp. 27-56.

[2] Robyt JF. Essentials of Carbohydrate Chemistry. Springer: New York 1998.

[3] James JA, Lee BH. Glucoamylase: Microbial sources, industrial applications and molecular biology-A review. J Food Biochem 1997; 21: 1-52.

[4] Hyun HH, Zeikus JG. Simultaneous and enhanced production of thermostable amylases and ethanol from starch by co-cultures of *Clostridium thermosulfurogenes* and *Clostridium thermohydrosulfuricum*. Appl Environ Microbiol 1985; 49: 1174-81.

[5] Zobel HF, Stephan AM. Starch: structure, analysis and application. In: Stephan AM, Ed. Food polysaccharide and their applications, Marcel Dekker Inc.: New York, Basel, Hongkong 1995.

[6] Hermansson AM, Svegmark K. Developments in the understanding of starch functionality. Trends Food Sci Technol 1996; 7: 345-53.

[7] Stephen AM, Churms SC. In: Stephan, AM, Ed. Food polysaccharide and their application, Marcel Dekker Inc. New York: Basel, Hongkong. 1995; pp. 1-20.

[8] Warren RAJ. Microbial hydrolysis of polysaccharides. Annu Rev Microbiol 1996; 50: 183-212.

[9] Oates CG. Towards and understand of starch granule structure and hydrolysis. TFST 1997; 8: 375-82.

[10] Arbige MV, Pitcher WH. Industrial Enzymology: a look towards the future. TIBTECH 1989; 7: 330-35.

[11] Vihinen M, Mantsala P. Microbial amylolytic enzymes. Crit Rev Biochem Mol Biol 1989; 24: 329-418.

[12] Kumar S, Satyanarayana T. Statistical optimization of a thermostable and neutral glucoamylase production by a thermophilic mould *Thermomucor indicae-seudaticae* in solid-state fermentation. World J Microbiol Biotechnol 2004a; 20: 895-902.

[13] Pandey A. Glucoamylase research. An overview. Starch/Starke 1995; 47: 439-45.

[14] Gerhartz W. Enzymes in industry-production and applications. V.C.H. Weinheim. 1990; pp. 113-48.

[15] Jensen B, Olsen J. Amylases and their industrial potential. Thermophilic Moulds in Biotechnology, In: Johri BN, Satyanarayana T, Olsen J. Eds. Kluwer Academic Publishers, Netherlands, 1999; pp. 115-37.

[16] Flemming ID. Starch and its derivatives 4th ed Radley, JA. Ed. Chapman and Hall Limited London. 1968; pp. 498-527.

[17] Reilly PJ. Protein engineering of glucoamylase to improve industrial properties. A Review. Starch/Starke 1999; 51: 269-74.

[18] Norman BE. A novel debranching enzyme for application in the glucose syrup industry. Starch/Starke 1982; 10: 340-6.

[19] Saha BC, Shen GJ, Srivastava KC, LeCureux LW, Zeikus JG. New thermostable α-amylase-like pullulanase from thermophilic *Bacillus* sp. 3183. Enzyme Microb Technol 1989; 11: 760-4.

[20] Satyanarayana T, Noorwez SM, Kumar S, Rao JLUM, Ezhilvannan M, Kaur P. Development of an ideal starch saccharification process using amylolytic enzymes from thermophiles. Biochem Soc Trans 2004; 32(2): 276-8.

[21] Antranikian G, Zablowski P, Gottschalk G. Conditions for the overproduction and excretion of thermostable α-amylase and pullulanase from *Clostridium thermohydrosulfuricum* DSM 567. Appl Microbiol Biotechnol 1987; 27: 75-81.

[22] Kim CH, Kim YS. Substrate specificity and detailed characterization of bifunctional amylase-pullulanase enzyme from *Bacillus circulans* F-2 having two different active sites on one polypeptide. Eur J Biochem 1995; 227: 687-93.

[23] Sonnleitner B, Fiechter A. Advantages of using thermophles in biotechnological process: expectations and reality. TIBTCH 1983; 1: 74-80.

[24] Kristjansson JK. Thermophilic organisms as sources of thermostable enzymes. TIBTECH 1989; 7: 349-53.

[25] Coolbear T, Daniel RM, Morgan HW. The enzymes from extreme thermophiles: Bacterial sources, thermo stabilities and industrial relevance. Adv Biochem Eng Biotechnol 1992; 45: 57-67.

[26] Adams MWW, Kelly RM. Finding and using hyperthermophilic enzymes. TIBTECH 1998; 16: 329-32.

[27] Janecek S. α-amylase family: molecular biology and evolution. Prog Biophys Mol Biol 1997; 67: 67-97.

[28] Henrissat BA. Classification of glycosyl hydrolases based on amino acid sequence similarities. Biochem J 1991; 280: 309-16.

[29] Kuriki T, Imanaka T. The concept of the α-amylase family: Structural similarity and common catalytic mechanism. J Biosci Bioeng 1999; 87: 557-65.

[30] Bhella RS, Altosaar I. Purification and some properties of the extracellular α-amylase from *Aspergillus awamori*. Can J Microbiol 1985; 31: 149-55.

[31] Olutiola PO. α-Amylolytic activity of *Aspergillus chevalieri* from mouldy maize seeds. Indian Phytopathol 1982; 35: 428-33.

[32] Khoo SL, Amirul AA, Kamaruzaman M, Nazalan N, Azizan MN. Purification and characterization of α-amylase from *Aspergillus flavus*. Folia Microbiol 1994; 39: 392-8.

[33] Michelena VV, Castillo FJ. Production of amylase by *Aspergillus foetidus* on rice flour medium and characterization of the enzyme. J Appl Bacteriol 1984; 56: 395-400.

[34] Domingues CM, Peralta RM. Production of amylase by soil fungi and partial biochemical characterization of amylase of a selected strain *(Aspergillus fumigatus)*. Can J Microbiol 1993; 39: 681-5.

[35] Alazard D, Baldensperger JF. Amylolytic enzymes from *Aspergillus hennebergi* (*A. niger* group): purification and characterization of amylases from solid and liquid cultures. Carbohydrate Res 1982; 107: 231-7.

[36] Bhumibhamon O. Production of amyloglucosidase by submerged culture. Thai J Agric Sci 1983; 16: 173-84.

[37] Yabuki M, Ono N, Hoshino K, Fukui S. Rapid induction of α-amylase by non-growing mycelia of *Aspergillus oryzae*. Appl Environ Microbiol 1977; 34: 1-6.

[38] Suganuma T, Tahara N, Kitahara K, Nagahama T, Inuzuka K. N-terminal sequence of amino acids and some properties of an acid stable α-amylase from citric acid Koji (*Aspergillus usamiivar*). Biosci Biotechnol Biochem 1996; 60: 177-9.

[39] Iefuji P, Chino M, Kato M, Iimura Y. Raw-starch digesting and thermostable α-amylase from the yeast *Cryptococcus* sp. S-2: purification, characterization, cloning and sequencing. Biochem J 1996; 318: 989-96.

[40] Prieto JA, Bort BR, Martinez J, Randez-Gil F, Buesa C, Sanz P. Purification and characterization of a new α-amylase of intermediate thermal stability from the yeast *Lipomyces kononenkoae*. Biochem Cell Biol 1995; 73: 41-9.

[41] Zenin CT, Park YK. Purification and characterization of acid α-amylase from *Paecilomyces* sp. J Ferment Technol 1983; 61: 109-14.

[42] Mishra RS, Maheshwari R. Amylases of the thermophilic fungus *Thermomyces lanuginosus*: Their purification, properties, action on starch and response to heat. J Biosci 1996; 21: 653-72.

[43] Nguyen QD, Judit M, Szabo R, Claeyssens M, Stals I, Hoschke A. Purification and characterization of amylolytic enzymes from thermophilic fungus *Thermomyces lanuginosus* strain ATTCC 34626. Enzyme Microb Technol 2002; 31: 345-52.

[44] Schellart JA, Visser FMW, Zandstva T, Middlehover WJ. Starch degradation by the mold *Trichoderma viride* I. The mechanism of degradation. Antonie van Leeuwenhock J Microbiol Serol 1976; 42: 229.

[45] Kumar P, Satyanarayana T. Optimization of culture variables for improving glucoamylase production by alginate-entrapped *Thermomucor indicaeseudaticae* using statistical methods. Bioresour Technol 2007b; 98: 1252-9.

[46] Walsh G. Industrial enzymes, an introduction In: Walsh G, Ed. Biochemistry and Biotechnology, Wiley J and Sons Co. Ltd: New York, 2002; pp. 393-454.

[47] Crabb W, Mitchinson C. Enzymes involved in the processing of starch to sugars. TIBTECH 1997; 15: 349-52.

[48] Rivera MH, Munguia AL, Soberon X, Rincon GS. α-Amylase from *Bacillus licheniformis* mutants near to the catalytic site: effects on hydrolytic and transglycosylation activity. Protein Eng 2003; 16: 505-14.

[49] Somkuti GA, Steinberg DH. Thermoacidophilic extracellular amylase of *Mucor pusillus*. Dev Ind Microbiol 1980; 21: 327-37.

[50] Niehaus F, Bertolldo C, Kahler M, Antranikian G. Extremophiles as a sources of novel enzymes for industrial application. Appl Microbiol Biotechnol 1999; 51: 711-29.

[51] Van der Maarel MJEC, Van der Veen B, Uitdehaag JCM, Leemhuis H, Dijhuizen L. Properties and applications of starch-converting enzymes of the α-amylase family. J Biotechnol 2002; 94: 137-55.

[52] Guzman-Maldonadao H, Paredes-Lopez O. Amylolytic enzymes and products derived from starch: a review. Crit Rev Food Sci Nutr 1995; 35: 373-403.

[53] Marc JEC, Maarel VD, Veen BVD, Uitdehaag JCM, Leemhuis H, Dijkhuizen L. Properties and applications of starch-converting enzymes of the α-amylase family. J Biotechnol 2002; 94: 137-55.

[54] Vielle C, Zeikus GJ. Hyperthermophilic Enzymes: sources, uses and molecular mechanisms for thermostability. Microbiol Mol Biol Rev 2001; 65: 1-43.

[55]   Pandey A, Nigam P, Soccol CR, Soccol VT, Singh D, Mohan R. Advances in microbial amylases. Biotechnol Appl Biochem 2000; 31: 135-52.

[56]   Nigam P, Singh D. Enzyme and microbial systems involved in starch processing. Enzyme Microb Technol 1995; 17: 770-78.

[57]   Fogarty WM, Kelly CT. Economic Microbiology. In: Microbial enzymes and bioconversions, (In: Rose AH.), Academic Press: London, 1980; Vol. 5, pp. 115-70.

[58]   Kelly CT, Collins BS, Fogarty WM, Doyle EM. High maltogenic α-amylases from the actinomycetes. Appl Microbiol Biotechnol 1993; 39: 599-603.

[59]   Van Ee JH, Rijswijk VWC, Bollier M. Enzymatic automated dish wash detergents. Chim Oggi 1992; 10: 21-4.

[60]   Kottwitz B, Upadek H, Carrer G. Application and benefits of enzymes in detergents. Chim Oggi 1994; 12: 21-4.

[61]   Svendsen A, Bisgard F. H PCT Patent Publication 1994; WO96/23874.

[62]   Bisgaard-Frantzen H, Borchert T, Svendsen A, Thellersen MH, Van DZP. PCY patent application 1995; WO95/10603 (to Novo-Nordisk A/S).

[63]   Bruinenberg PM, Hulst AC, Faber A, Voogd RH. A process for surface sizing or coating of paper. European patent application 1996; EP 0690 170 A1.

[64]   Hamer RJ. Enzymes in the baking industry. In: Enzymes in food processing Tucker GA, Woods LFJ. Eds, Blackie Academic and Professional: Galsgow 1995; pp. 190-222.

[65]   Si JQ. Enzymes, baking, bread making. John Wiley & Sons Inc.1999; 2:947-58.

[66]   Hebeda RE, Bowles LK, Teague WM. Developments in enzymes for retarding staling of baked goods. Cereal Foods World 1990; 35(5): 453-7.

[67]   Hebeda RE, Bowles LK, Teague WM. Use of intermediate temperature stability enzymes for retarding staling in baked goods. Cereal Foods World 1991; 36: 619-24.

[68]   Van Dam HW, Hille JDR. Yeast and enzymes in bread making. Cereal Foods World 1992; 37(2): 245-52.

[69]   Kulp K, Ponte JG. Staling of white pan bread: fundamental causes. CRC Crit Rev Food Sci Nutr 1981; 15: 1-48.

[70]   Spendler T, Jorgensen O. Use of a branching enzyme in baking. Patent application 1997; WO97/41736.

[71]   Destefains VA, Turner EW. Modified enzymes system to inhibit bread firming method for preparing same and use of same in bread and other bakery products. Patent Application US4299848.

[72]   Cole MS. Antistaling baking composition, Patent application 1982; US 4320151.

[73]   Carroll JO, Boyce COL, Wong TM, Starace CA. Bread antistaling method. US Patent Application 1987; No. 4654216.

[74]   De Moraes LMP, Astolfi-Filho S, Oliver SG. Development of yeast strains for the efficient utilisation of starch: evaluation of constructs that express α-amylase and glucoamylase separately or as bifunctional fusion proteins. Appl Microbiol Biotechnol 1995; 3: 1067-76.

[75]   Coombs J. Sugarcane as an energy crop. Biotechnol Gen Eng Rev 1984; 1: 311-45.

[76]   Huang SY, Chen JC. Ethanol production in simultaneous saccharification and fermentation of cellulose with temperature profiling. J Ferment Technol 1988; 66: 509-16.

[77]   Szczodark J, Tagonski Z. Simultaneous saccharification and fermentation of cellulose: effect of ethanol and cellulases on particular stages. Acta Biotechnol 1989; 9: 555-64.

[78]   Kumar S, Satyanarayana T. Microbial glucoamylases: Characteristics and applications. Crit Rev Biotechnol 2009; 29: 225-55.

[79]   Ford C. Improving operating performance of glucoamylase by mutagenesis. Curr Opin Biotechnol 1999; 10: 352-7.

[80]   Manjunath P, Shenoy BC, Raghavendra Rao MR. Fungal glucoamylase. J Appl Biochem 1983; 5: 235-60.

[81]   Gibbs PA, Seviour RJ, Schmid F. Growth of filamentous fungi in submerged culture: problems and possible solutions. Crit Rev Biotechnol 2000; 20: 17-48.

[82]   Ramadas M, Holst O, Mattiasson B. Production of amyloglucosidase by *Aspergillus niger* under different cultivation regimens. World J Microbiol Biotechnol 1996; 12: 267-71.

[83]   Kumar S, Satyanarayana T. Production of thermostable and neutral glucoamylase by a thermophilic mould *Thermomucor indicae-seudaticae* in solid-state fermentation. Indian J Microbiol 2004b; 44: 53-7.

[84]   Pedersen H, Beyer M, Nielson J. Glucoamylase production in batch, chemostat and fed batch cultivation by an industrial strain of *Aspergillus niger*. Appl Microbiol Biotechnol 2000; 53: 272-7.

[85]   Merico A, Capitanio D, Vigentini I, Ranzi BM, Compagno C. How physiological and cultural conditions influence heterologous protein production in *Kluyveromyces lactis*. J Biotechnol 2004; 109: 139-46.

[86]   Ganzlin M, Rinas U. In-depth analysis of the *Aspergillus niger* glucoamylase (glaA) promoter performance using high-throughput screening and controlled bioreactor cultivation techniques. J Biotechnol 2008; 135: 266-71.

[87]   Rao BV, Sastri NVS, Subba Rao PV. A thermostable glucoamylase from the thermophilic fungus *Thermomyces lanuginosus*. Biochem J 1979; 193: 379-87.

[88]   Aalbaek T, Reeslev M, Jensen B, Eriksen SH. Acid protease and formation of multiple forms of glucoamylase in batch and continuous cultures of *Aspergillus niger*. Enzyme Microb Technol 2002; 30: 410-5.

[89]   Riaz M, Perveen R, Javed MR, Nadeem H, Rashid MH. Kinetic and thermodynamic properties of novel glucoamylase from *Humicola* sp. Enzyme Microb Technol 2007; 41: 558-64.

[90]   Bhatti HN, Rashid MH, Nawaz R, Asgher M, Perveen R, Jabbar A. Purification and characterization of a novel glucoamylase from *Fusarium solani*. Food Chem 2007b; 103: 338-43.

[91]   Chen J, Li DC, Zhang YQ, Zhou QX. Purification and characterization of a thermostable glucoamylase from *Chaetomium thermophilum*. J Gen Appl Microbiol 2005; 51: 175-81.

[92]   Aquino AC, Jorge JA, Terenzi HF, Polizeli MLTM. Thermostable glucose-tolerant glucoamylase produced by the fungus *Scytalidium thermophilum*. Folia Microbiol 2001; 46: 11-6.

[93]   Cereia M, Terenzi HF, Jorge JA, Greene LJ, Rosa JC, Polizeli MLTM. Glucoamylase activity from the thermophilic fungus *Scytalidium thermophilum*: biochemical and regulatory properties. J Basic Microbiol 2000; 40: 83-92.

[94]   Nguyen QD, Rezessy-Szabo JM, Claeysssens M, Stals I, Hoschke A. Purification and characterization of amylolytic enzymes from thermophilic fungus *Thermomyces lanuginosus* strain ATCC 34626. Enzyme Microb Technol 2002; 31: 345-52.

[95]   Feng B, Hu W, Ma BP, *et al.* Purification, characterization, and substrate specificity of a glucoamylase with steroidal saponin-rhamnosidase activity from *Curvularia lunata*. Appl Microbiol Biotechnol 2007a; 76: 1329-38.

[96]   Thorsen TS, Johnsen AH, Josefsen K, Jensen B. Identification and characterization of glucoamylase from the fungus *Thermomyces lanuginosus*. Biochem Biophys Acta 2006; 1764: 671-6.

[97]   Ono K, Shigeta S, Oka S. Effective purification of glucoamylase in Koji, a solid culture of *Aspergillus oryzae* on steamed rice, by affinity chromatography using an immobilized acarbose (BAY g-542). Agric Biol Chem 1988; 52: 1707-14.

[98]   Kumar S, Satyanarayana T. Purification and kinetics of a raw starch-hydrolyzing, termostable, and neutral glucoamylase of the thermophilic mold *Thermomucor indicae-seudaticae*. Biotechnol Prog 2003; 19: 936-44.

[99]   Norouzian D, Rostami K, Nouri ID, Saleh M. Subsite mapping of purified glucoamylases I, II, III produced by *Arthrobotrys amerospora* ATCC34468. World J Microbiol Biotechnol 2000; 16: 155-61.

[100] Michelin M, Ruller R, Ward RJ, *et al.* Purification and biochemical characterization of a thermostable extracellular glucoamylase produced by the thermotolerant fungus *Paecilomyces variotii*. J Ind Microbiol Biotechnol 2008; 35: 17-25.

[101] Synowiecki J. The use of starch processing enzymes in the food industry. In Polaina J, Mac Cabe AP, Eds. Industrial Enzymes, Structure, Function and Applications. Netherlands: Springer. 2007; pp. 19-34.

[102] Crabb WD, Shetty JK. Comodity scale production of sugars from starches. Curr Opin Microbiol 1999; 2: 252-6.

[103] Murai T, Ueda M, Shibasaki Y, Kamasawa N, Osumi M, Imanaka T, Tanaka A. Development of an arming yeast strain for efficient utilization of starch by co-display of sequential amylolytic enzymes on the cell surface. Appl Microbiol Biotechnol 1999; 51: 65-70.

[104] Altintas MM, Ulgen K, Kirdar B, Onsan ZI, Oliver SG. Improvement of ethanol production from starch by recombinant yeast through manipulation of environmental factors. Enzyme Microb Technol 2002; 31: 640-7.

[105] Alkorta I, Garbisu C, Liama MJ, Serra JS. Industrial applications of pectic enzymes: a review. Process Biochem 1998; 33: 21-8.

[106] Baracat-Pereira MC, Vanetti MCD, Araujo EFD, Silva DO. Partial characterization of *Aspergillus* fumigatus polygalacturonases for the degumming of natural fibres. J Ind Microbiol 1993; 11: 139-42.

[107] Henriksson G, Akin DE, Solmezynski D, Erikson KEL. Production of highly efficient enzymes for flax retting by *Rhizomucor pusillus*. J Biotechnol 1999; 68: 115-23.

[108] Voragen A, Wolters H, Verdonschot-Kroef T, Rombouts FM, Pilnik W. Effect of juice-releasing enzymes on juice quality. In: International Fruit Juice Symposium Juris Durck Verlag, Zurich, Switzerland, 1986; pp. 453-62.

[109] Kaur G, Kumar S, Satyanarayana T. Production, characterization and application of a thermostable polygalacturonase of a thermophilic mould *Sporotrichum thermophile* Apinis. Bioresour Technol 2004; 94: 239-43.

[110] Jenson B, Olsen J. Miscellaneous enzymes. In: Johri BN, Satyanarayana T, Olsen J. Eds. Thermophilic Moulds in Biotechnology. Kluwer Academic Publishers: Netherlands 1999; pp. 245-63.

[111] Endo A. Studies on pectinolytic enzymes of molds. Part VII. Turbidometry of apple juice clarification and its application to determination of enzyme activity. Agric Biol Chem 1964; 28: 234-8.

[112] Yamasaki M, Yasui T, Arima K. Pectic enzymes in the clarification of apple juice. Part I. Study on the clarification reaction in a simplified model. Agric Biol Chem 1964; 28: 779-87.

[113] Grassin C, Fauquembergue P. Fruit Juices. In: Godfrey T, West S, Eds. Industrial Enzymology. MacMillan Press: London, 1996; pp. 226-64.

[114] Sakai T, Sakamoto T, Vandamne EJ. Pectin, pectinase and protopectinase: production, properties and applications. Adv Appl Microbiol 1993; 39: 213-94.

[115] Zhou P. *Application of Pectinex Smash XXL and Pectinex Ultra SP-L in Carrot Juice Processing.* Sci Technol Food Ind 2003; 9: 62-3.

[116] Godfrey T, West S. Introduction to industrial enzymology. In: Godfrey T, West S, Eds. Industrial Enzymology, 2nd ed. Stockholm Press, New York, 1996; pp. 1-17.

[117] Carr JG. Tea, coffee and cocoa. In: Wood BJB, Ed. Microbiology of Fermented Foods, Elsevier Applied Science, London, 1985; vol. II. pp. 133-54.

[118] Acuna-Arguelles, ME, Gutierrez-Rajas M, Viniegra-Gonzalez G Favela-Toress E. Production and properties of three pectinolytic activities produced by *A. niger* in submerged and solid state fermentation. Appl Microbiol Biotechnol 1995; 43: 808-14.

[119] Al-Obaidi ZS, Aziz GM, Al-Bakir AY. Screening of fungal strains for polygalacturonase production. J Agric Water Resour Res 1987; 6: 125-82.

[120] Borin MDF, Said S, Fonseca MJV. Purification and biochemical characterization of an extracellular endopolygalacturonase from *Penicillium frequentans*. J Agric Food Chem 1996; 44: 1616-20.

[121] Marcus L, Barash I, Sneh B, Koltin Y, Finker A. Purification and characterization of pectolytic enzymes produced by virulent and hypovirulent isolates of *Rhizoctonia solani* KUHN. Physiol Mol Plant Pathol 1986; 29: 325-36.

[122] Channe PS, Shewal JG. Pectinase production by *Sclerotium rolfsii*: Effect of culture conditions. Folia Microbiol 1995; 40: 111-7.

[123] Ahmad T, Rasool S, Sarwar M, Haq AU, Hasan ZU. Effect of microbial phytase produced from a fungus *Aspergillus niger* on bioavailability of phosphorus and calcium in broiler chickens. Anim Feed Sci Technol 2000; 83: 103-14.

[124] Kim DS, Godber JS, Kim HR. Culture conditions for a new phytase-producing fungus. Biotechnol Lett 1999; 21: 1077-81.

[125] Ebune A, Alasheh S, Duvnjak Z. Production of phytase during solid-state fermentation using *Aspergillus-ficuum* NRRL-3135 in canola-meal. Bioresour Technol 1995; 53: 7-12.

[126] Alasheh S, Duvnjak Z. Characteristics of phytase produced by *Aspergillus carbonarius* Nrc-401121 in canola-meal. Acta Biotechnol 1994a; 14: 223-33.

[127] Alasheh S, Duvnjak Z. Effect of glucose-concentration on the biomass and phytase productions and the reduction of the phytic acid content in canola-meal by *Aspergillus carbonarius* during a solid-state fermentation process. Biotechnol Prog 1994b; 10: 353-9.

[128] Alasheh S, Duvnjak Z. Phytase production and decrease of phytic acid content in canola-meal by *Aspergillus-Carbonarius* in solid-state fermentation. World J Microbiol Biotechnol 1995; 11: 228-31.

[129] Hardy RW, Shearer KD. Effect of dietary calcium phosphate and zinc supplementation on whole body zinc concentration of rainbow trout (*Salmo gairdneri*). Can J Fish Aquat Sci 1985; 42: 181-4.

[130] Correll DL. Phosphorus: a rate limiting nutrient in surface waters. Poult Sci 1999; 78: 674-82.

[131] Rodehutscord M, Pfeffer E. Effects of supplemental microbial phytase on phosphorus digestibility and utilization in rainbow trout (*Oncorhynchus mykiss*). Water Sci Technol 1995; 31: 143-7.

[132] Baruah K, Sahu NP, Pal AK, Debnath D. Dietary phytase: an ideal approach for a cost effective and low-polluting aquafeed. NAGA, WorldFish Center Q 2004; 27: 15-9.

[133] Cao L, Wang W, Yang C, *et al.* Application of microbial phytase in fish feed. Enzyme Microb Technol 2007; 40(4): 497-507.

[134] Yu FN, Wang DZ. The effects of supplemental phytase on growth and the utilization of phosphorus by crucian carp *Carassius carassius*. J Fish Sci Chin 2000; 7: 106-9.

[135] Fu SJ, Sun JY. Application of neutral phytase in aquaculture. J Chin Feed 2005; 2: 25-7.

[136] Lei XG, Stahl CH. Nutritional benefits of phytase and dietary determinants of its efficancy. J Appl Anim Res 2000; 17: 97-112.

[137] Mitchell DB, Vogel K, Weimann BJ, Pasamontes L, van Loon APGM. The phytase subfamily of histidine acid phosphatases: isolation of genes for two novel phytases from the fungi *Aspergillus terreus* and *Myceliophthora thermophile*. Microbiology 1997; 143: 245-52.

[138] Wodzinski RJ, Ullah AHJ. Phytase. Adv Appl Microbiol 1996; 42: 263-310.

[139] Vats P, Banerjee UC. Biochemical characterisation of extracellular phytase (myo-inositol hexakisphosphate phosphohydrolase) from a hyper-producing strain of *Aspergillus niger* van Teighem. J Ind Microbiol Biotechnol 2005; 32(4): 141-7.

[140] Greiner R, Konietzny U. Phytase for food application. Food Technol Biotechnol 2006; 44(2): 125-40.

[141] Berridge MJ, Irvine RF. Inositol phosphates & cell signalling. Nature 1989; 341: 197-205.

[142] Maga JA. Phytate: Its chemistry, occurrence, food interactions, nutritional significance, & methods of analysis. J Agri Food Chem 1982; 30: 1-9.

[143] Harland BF, Morris ER. Phytate: A good or a bad food component. Nutr Res 1995; 15: 733-54.

[144] Mullaney EJ, Daly CB, Ullah AHJ. Advances in phytases research. Adv Appl Microbiol 2000; 47: 157-99.

[145] Satyanarayana T, Vohra A, Kaur P. Phytases in animal productivity and environmental management. Productivity 2004; 44: 542-8.

[146] Nelson TS. The utilization of phytate phosphorus by poultry. Poult Sci 1967; 46: 862-71.

[147] Ciofalo V, Barton N, Kertz K, Biaird J, Cook M, Shahnahan D. Safety evaluation of a phytase, expressed in *Schizosaccharomyces pombe*, intended for use in animal feed. Regul Toxicol Pharmacol 2003; 37: 286-92.

[148] Chantasartrasamee K, Ayuthaya DIN, Intarareugsorn S, Dharmsthiti S. Phytase activity from *Aspergillus oryzae* AK9 cultivated on solid state soybean meal medium. Process Biochem 2005; 40: 2285-9.

[149] Bogar B, Szakacs G, Linden JC, Pandey A, Tengerdy RP. Optimization of phytase production by solid substrate fermentation. J Ind Microbiol Biotechnol 2003; 30: 183-9.

[150] Singh B, Kaur P, Satyanarayana T. Fungal phytases for improving the nutritional status of foods and combating environmental phosphorus pollution. In: Chauhan AK, Verma A, Eds. Microbes: Health and Environment. IK International Publishers, New Delhi, India, 2006; pp. 289-326.

[151] McCoy M. Enzymes emerge as big AG feed supplements. Chem Eng News 1998; 4: 29-30.

[152] Nampoothiri KM, Tomes GJ, Roopesh K, *et al.* Thermostable phytase production by *Thermoascus aurantiacus* in submerged fermentation. Appl Biochem Biotechnol 2004; 118 (1–3):205-14.

[153] Singh B, Satyanarayana T. Phytase production by a thermophilic mould *Sporotrichum thermophile* in solid state fermentation and its application in dephytinization of sesame oil cake. Appl Biochem Biotechnol 2006a; 133(3): 239-50.

[154] Singh B, Satyanarayana T. A marked enhancement in phytase production by a thermophilic mould *Sporotrichum thermophile* using statistical designs in a cost-effective cane molasses medium. J Appl Microbiol 2006b; 101(2): 344-52.

[155] Singh B, Satyanarayana T. Improved phytase production by a thermophilic mould *Sporotrichum thermophile* in submerged fermentation due to statistical optimization. Bioresour Technol 2008a; 99(4): 824-30.

[156] Singh B, Satyanarayana T, Phytase production by *Sporotrichum thermophile* in solid state fermentation and its applications. Bioresour Technol 2008b; 99(8): 2824-30.

[157] Singh B, Satyanarayana T. Phytase production by *Sporotrichum thermophile* in a cost-effective cane molasses medium in submerged fermentation and its application in bread. J Appl Microbiol Bioresour Technol 2008; 105: 1858-65.

[158] Machius M, Wiegand G, Humber R. Crystal structure of calcium-depleted *Bacillus licheniformis* α-amylase at 2.2 A resolution. J Mol Biol 1995; 246: 545-59.

[159] Swift HJ, Brady L, Derewenda ZS, *et al.* Structure and molecular model refinement of *Aspergillus oryzae* (TAKA) α-amylase: an application of the simulated-annealing method. Acta Crystallogr B 1991; 47: 535-44.

[160] Aehle W, Arfman N, Koenhen E, Scholte M, Laan Van Der JM. Improvement of *Bacillus* α-amylase for industrial applications by protein engineering. American Oil Chemical Society, 1995, 86th Annual Meeting, San Antonio.

[161] Hagihara H, Igarashi K, Hayashi Y, *et al.* Novel α-Amylase that is highly resistant to chelating reagents and chemical oxidants from the alkaliphilic *Bacillus* isolate KSM-K38. Appl Environ Microbiol 2001; 67(4): 1744-50.

[162] Turner NJ. Directed evolution of enzymes for applied biocatallysis. Trends Biotechnol 2003; 21: 474-7.

[163] Weber FJ, Tramper J, Rinzema A. A simplified material and energy balance approach for process development and scale-up of *Coniothyrium minitans* conidia production by solid-state cultivation in a packed bed reactor. Biotechnol Bioeng 1999; 65: 447-58.

[164] Abelson PH, A potential phosphate crisis. Science 1999; 283: 2015.

CHAPTER 2

Fungal Lignocellulolytic Enzymes: Applications in Biodegradation and Bioconversion

Juana Pérez* and Aurelio Moraleda-Muñoz

Departamento de Microbiología. Facultad de Ciencias. Instituto de Biotecnología. Avda. Fuentenueva s/n. Universidad de Granada. E-18071. Granada. Spain

Abstract: Lignocellulosic materials as industrial, agricultural, and forest residues account for the majority of the total renewable biomass present on Earth. Some fungi are equipped with potent enzymatic systems involved in the hydrolysis or oxidation of these biopolymers. Herein, we provide an update of lignocellulose biodegradation processes and the main biotechnological applications of lignocellulolytic fungi or their enzymes in biotransformation and biodegradation of wastes, and in the conversion of biomass into valuable products.

INTRODUCTION

The term lignocellulosic material (LCM) refers to the plant material, produced *via* photosynthesis and it is composed of three different biological polymers: cellulose, hemicellulose, and lignin, which are strongly intermeshed and bonded by covalent cross-linkages and non covalent forces [1]. It is important to notice that, although lignocellulosics are included in the term "biomass", this word has broader implications and also includes other non-photosynthetic substances such as animal tissues or bones. Lignocellulosic biomass accounts for the majority of the total renewable resources present on Earth, with an annual production of approximately 2×10^{11} tons [2, 3]. The main sources of lignocelluloses include wood, grasses, water plants, agricultural, industrial and forest residues, municipal solid wastes, and dedicated bioenergy crops.

The wood produced yearly by terrestrial plants is around 1.3×10^{10} tons [4] and the global wood consumption is around 3.5×10^9 m^3 per year [5]. Most of this exploitation is for pulp and paper industries, building materials, and production of fuels. The worldwide agricultural production data are collected in the FAO webpage (http://faostat.fao.org/). Millions of tons of cereals, oil seed, pulse crops, and different type of other crops are produced annually for essential demands such as animal and human feed, or basic materials, and for secondary demands such as fuels and chemicals. This enormous production generates also a large amount of lignocellulosic residues (for a review with agricultural residues data see [6]). The bioconversion of lignocellulose biomass to valuable products is being more appealing day after day due not only to the advantages that suppose its abundance and its low cost, but also the growing demand for renewable energy and sustainable biofuels.

However, the direct use of lignocellulose biomass as feedstock is difficult (especially for the microorganisms or enzymes based-technologies) owing to its complex structure. Consequently, it is important to understand the pathways developed by microorganisms to biodegrade or modify lignocellulose, in order to design and improve strategies to make the processes feasible and economically viable.

In nature, the degradation of lignocellulose biomass is accomplished mainly by fungi, being the most rapid degraders the wood rotting basidiomycetes, which are classified as white-rot fungi (WRF) and brown-rot fungi (BRF). The most numerous and efficient wood rotting are the WRF, which own this name to their ability for preferential degradation of lignin polymer giving rise to a cellulose-enriched white material, that can be observed as white regions of the mottled rot and in the pockets of a white pocket rot [7]. The brown-rot fungi, on the contrary, degrade cellulose and hemicellulose along with a partial modification of lignin, resulting in a brown residual matter, mainly composed by oxidized lignin. It is out of the scope of this chapter to describe the different wood decay patterns caused by fungi or the type of fungi that produce them (for a comprenhensive review see Martinez *et al.* [5]).

*Address correspondence to Juana Perez: Departamento de Microbiología. Facultad de Ciencias. Instituto de Biotecnología. Avda. Fuentenueva s/n. Universidad de Granada. E-18071. Granada. Spain; E-mail: jptorres@ugr.es

Lignocellulolytic microorganisms can fragment the macromolecules that compose lignocellulose by using two different mechanisms: i) the hydrolytic cellulose and hemicellulose degrading systems, composed by a battery of hydrolases with different specificities; and ii) the battery of non-specific oxidative enzymes involved in the biological lignin biodegradation process. Indeed, the low substrate specificity confers to the ligninolytic enzymes the capacity to degrade a broad variety of recalcitrant compounds and, consequently, the biotechnological approaches have been extended since the historical discovery of the first lignin degrading enzyme in 1983 by Tien and Kirk [8].

Many research groups around the world are currently studying the use of these filamentous fungi or their enzymes in a variety of biotechnological processes such as biopulping, biobleaching or elimination of pitch in the pulp and paper industries, pretreatments of lignocellulose biomass for production of biofuels, soil bioremediation, improvement of compost, degradation and/or decolourization of effluents, use as additives in the food, cosmetic and textile industries, or even in organic synthesis. The main goal of this review is to give a general overview of the composition of LCMs in order to understand the basis of biodegradation mechanisms. Furthermore, the understanding of catalytic mechanisms of the lignocellulolytic enzymes is required for a successful integration of these fungi or their enzymes in industrial and biotechnological processes.

COMPOSITION, STRUCTURE AND DISTRIBUTION OF THE LIGNOCELLULOSICS

The comprehension of the structure and composition of the LCMs is essential for understanding the bioconversion mechanisms. The major components of lignocellulosic materials are cellulose, hemicellulose, and lignin. Cellulose and hemicellulose are macromolecules composed by different sugars, whereas lignin is an amorphous aromatic and highly complex three-dimensional polymer, synthesized from three phenylpropanoid precursors. The composition and percentages of these macromolecules vary from one plant species to another [6,9], and it is also dependent on age, stage of growth, and other conditions [1].

Cellulose represents about 45-50% of the dry weight of wood, but its percentage varies widely in other LCMs and can range from 6% to 90% [6]. This polysaccharide is a linear and highly ordered macromolecule composed by D-glucose subunits, bound together by β-1,4 glucoside linkages forming the dimer cellobiose. These repetitive units form long chains known as "elemental fibrils" which are linked together by van der Waals forces and hydrogen bonds to give "microfibrils". A number of microfibrils form "macrofibrils". The major part of cellulose appears in a stable crystalline form, although a small fraction of non organized cellulose chains, called amorphous cellulose, can also be found. In the plant cell walls, cellulose microfibrils and hemicellulose chains are covered by lignin matrix which protects them against biodegradation (Fig. **1**).

Hemicelluloses are heterogeneous branched polysaccharides, with a lower molecular weight than cellulose due to its smaller degree of polymerization. In contrast to cellulose, these macromolecules do not form aggregates and are therefore more easily hydrolysable polymers. It makes up 25-30% of wood dry weight, but its proportion is variable in others LCMs ranging from 4% to 85% [6]. Three hexoses (D-glucose, D-mannose, and D-galactose), two pentoses (D-xylose and L-arabinose), glucuronic and galacturonic acids, and acetylated sugars are the main constituents of hemicelluloses.

The most common hemicelluloses in hardwood are glucuronoxylans (xylans) which are composed mainly by xylose, which are more or less acetylated depending on LCM species belong to angiosperms or gymnosperms. Additionally, this basic structure holds different percentages of mannose, galactose or arabinose and methylglucuronic acids. In softwoods, the predominant structures are glucomannans, frequently combined with galactose to form galactoglucomannans (for detailed structures see [1]). The hemicellulose chains surround the cellulose fibres and act as a linkage between cellulose and lignin. The hemicellulose-cellulose interconnections are achieved by non-covalently linkage, mostly by hydrogen bonds. However, the hemicellulose layers are covalently linked to lignin mainly by diferulic bridges and ester linkage between lignin and glucuronic acid attached to xylans [10]. These lignin-hemicellulose covalent crosslinks provide strength, rigidity and hydrophobicity to the cell walls.

Lignin is an abundant carbon source on Earth, second only to cellulose in terms of biomass. Its complex structure is formed by dehydrogenative polymerization of three primary hydroxycinnamyl alcohols (monolignols): coniferyl alcohol (guaiacyl propanol, G), coumaryl alcohol (*p*-hydroxyphenyl propanol, H), and sinapyl alcohol (syringyl propanol, S) (Fig. **2**). The H:G:S ratio varies between different vascular plant groups [5, 7].

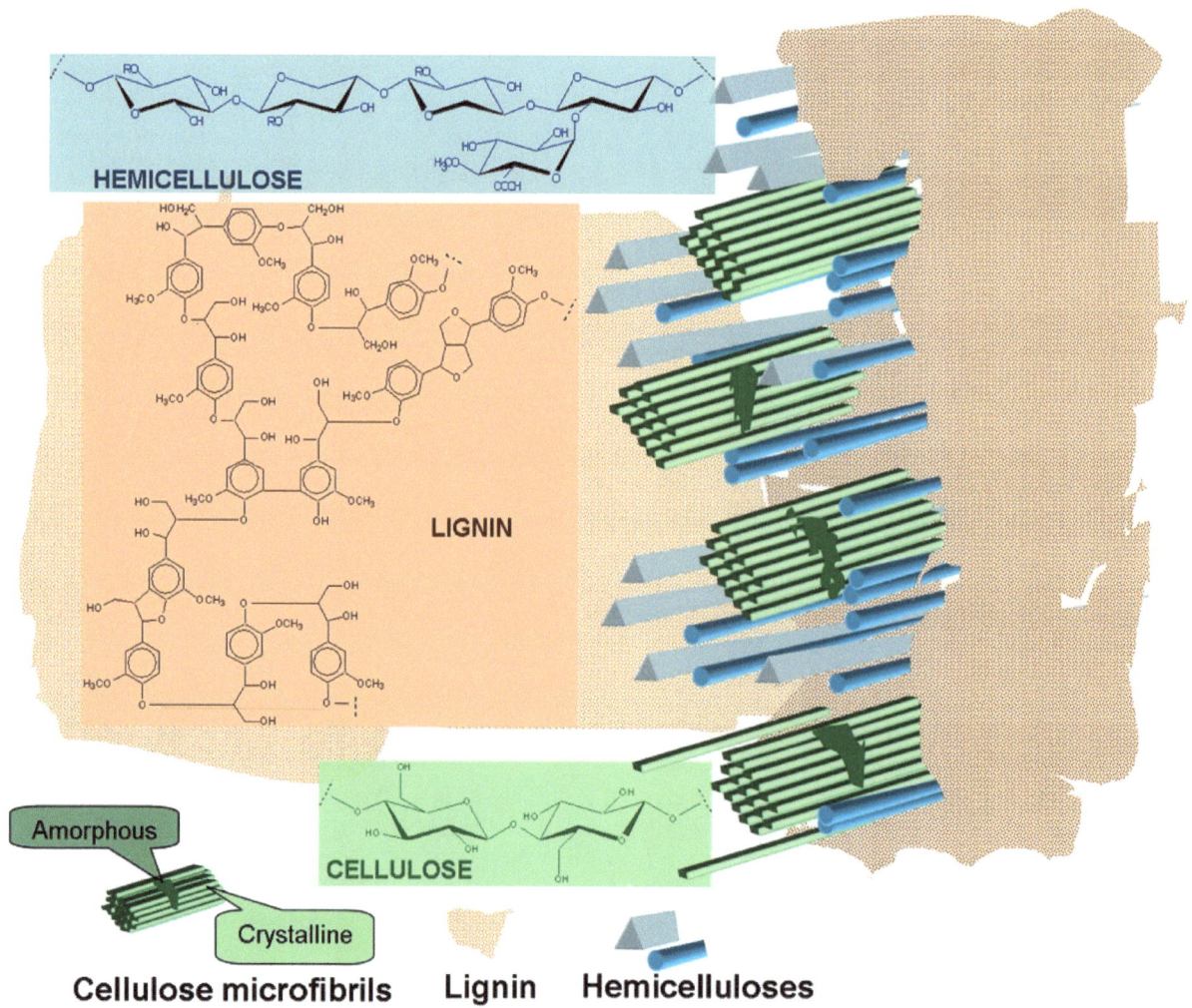

Figure 1: Distribution of cellulose, hemicellulose, and lignin in the cell walls. Cellulose fibrils and hemicellulose chains are embedded in the lignin matrix.

Lignin monomeric precursors (hydroxycinnamyl alcohols)

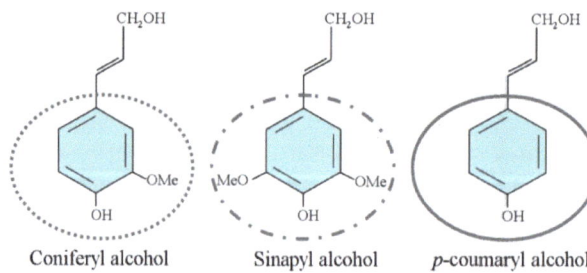

Corresponding structural units in lignin:

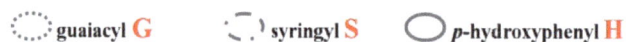

Figure 2: Lignin is formed *via* radicals by enzymatic oxidative coupling of the three phenylpropane monomers.

These three monolignols are originated in plants *via* the phenylpropanoid pathway [11]. These lignin building blocks bound together by different type of linkages such as ether or carbon-carbon bonds, being the most common the arylglycerol-β-aryl ether β-O-4 linkage (Fig. **3**), which makes about a half of the total structures [7]. The resulting macromolecule is an amorphous heteropolymer, non water soluble, and optically inactive, which confers support and impermeability to the cell walls and it is highly recalcitrant to breakdown by most microorganisms.

The different chemical composition and structure of cellulose, hemicellulose and lignin require different biodegradation systems. In the three cases, however, the degradation processes have to take place exocellularly due to the insolubility and complexity of the substrates.

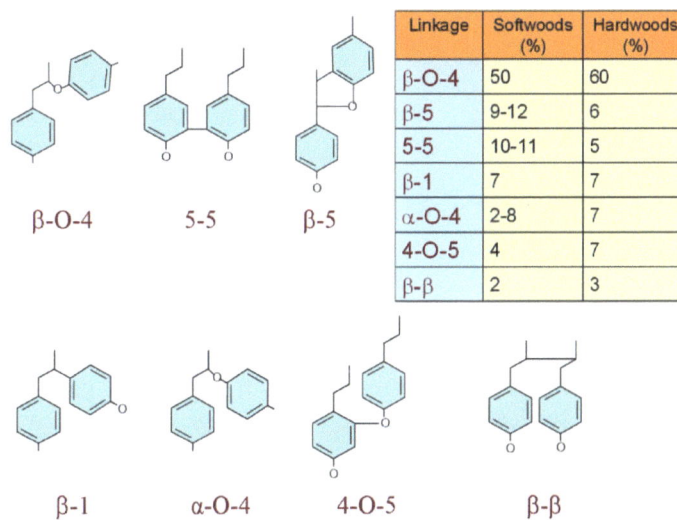

Linkage	Softwoods (%)	Hardwoods (%)
β-O-4	50	60
β-5	9-12	6
5-5	10-11	5
β-1	7	7
α-O-4	2-8	7
4-O-5	4	7
β-β	2	3

Figure 3: Principal modes of linkages between monomeric phenylpropane units and their relative abundance in softwood and hardwood lignins.

HYDROLYTIC BIODEGRADING SYSTEMS

Cellulose Biodegradation

The cellulose hydrolysing systems (also called cellulase enzymes systems) are formed by different groups of enzymes with different specificities working together. Cellulases are able to hydrolyse the β-1,4-glucosidic linkage of cellulose. Conventionally they have been divided in three classes: endoglucanases, cellobiohydrolases (exoglucanases) and β-glucosidases. Endoglucanases (1,4-β-D-glucan-4-glucanohydrolases, EGs) catalyze cleavage of internal bonds of the cellulose chain, preferably in the amorphous regions, generating oligosaccharides of various lengths. The exoglucanases cellobiohydrolases (1,4-β-D-glucan glucanohydrolases, CBHs) attack the reducing or non reducing ends of the cellulose chains releasing the disaccharide cellobiose. Finally, β-glucosidases (β-glucoside glucohydrolases) hydrolyze cellobiose to two glucose molecules [12]. All the enzymes that are part of the cellulase systems exhibit a high degree of synergism and act in a coordinate manner to efficiently hydrolyze cellulose.

During many years it was maintained that the hydrolysis and utilization of cellulose were carried out exclusively by bacteria and fungi. In fact, it is well documented that more than 5% of cellulose is degraded in nature by bacteria under anaerobic conditions. The bacterial anaerobic cellulose-degrading system is very different from that of aerobic microorganisms. Thus, some anaerobic bacteria, such as *Clostridium thermocellum*, show the enzymes involved in cellulose biodegradation organized into large functional entities termed cellulosomes. The organization of these enzymes into cellulosome concentrates and positions them in such a manner that promotes synergism among catalytic units [13]. Now, we know that other organisms, such as termites or crayfish, also produce cellulases. Fungi with cellulolytic capability are distributed from the primitive anaerobic Chytridomycetes to the advanced subdivision of Ascomycetes, Basidiomycetes or Deuteromycetes. Many species belonging to these three groups have been widely studied not only for their cellulolytic capabilities, but also for their ability to degrade wood. The most well-known among them are the Ascomycete *Chaemotium*, the Basidiomycetes *Coriolus*, *Phanerochaete* and

Poria and the Deuteromycetes *Aspergillus, Cladosporium, Fusarium, Geotrichum, Penicillium* or *Trichoderma* and the Deuteromycete *Humicola insolens* [14]. Among all, the cellulase system of the mesophilic fungi *Trichoderma reesei* and *Phanerochaete chrysosporium* and the thermophilic fungus *H. insolens* are the most thoroughly studied. *T. reesei* produces two cellobiohydrolases, five endoglucanases and two β-glucosidases. *P. chrysosporium* secretes seven cellobiohydrolases, one endoglucanase and several β-glucosidases. The cellulase system of *H. insolens* is homologous to *T. reesei* system and also contains two cellobiohydrolases and five endoglucanases. The cellulases of the thermophilic fungi are thermostable and they show optimal activities to very high temperatures: 55-80 °C for EGs, 50-75 °C for CBHs, and 35-71 °C for β-glucosidases [14-16]. In general, the EGs of these non complexes cellulase systems have two structural domains: the catalytic and the union domains. Their molecular masses vary from 25 to 50 kDa and they show the optimal activity at acidic pHs. CBHs act synergistically with EGs to solubilize high molecular weight cellulose. Microbial cellulose degradation is very important not only because it is essential to close the Earth carbon cycle, but also because the great implication in biotechnology of cellulosic biomass. Cellulosic materials are attractive because of relatively low cost, large-scale availability and environmental friendly production. However, the main obstacle to more extensive utilization is the absence of low-cost technology for defeating the recalcitrance of these materials [17].

Hemicelluloses Biodegradation

The complex structure of hemicelluloses requires the concerted action of a battery of enzymes. The hydrolytic breakdown releases mainly monomeric sugars and acetic acids. The degradation of xylan, the major constituent of hemicelluloses, in herbaceous plants (arabinoxylan) and hardwood (glucuronoxylan), requires the cooperative action of a variety of hydrolytic enzymes. First, the endo-1,4-β-xylanases generate oligosaccharides from the hydrolysis of the xylan backbone. Second, the xylan 1,4-β-xylosidase works on these xylan oligosaccharides to produce xylose [18]. The complexity of the branched hemicellulose molecules requires the hydrolytic action of accessory enzymes.

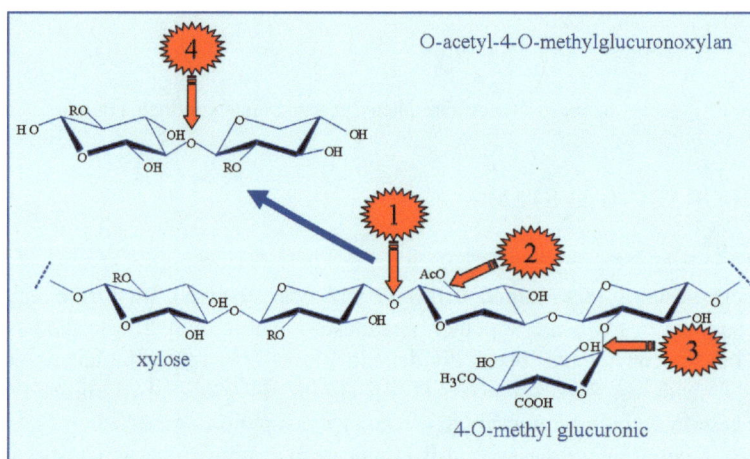

Figure 4: Structure of the most frequent hemicellulose in angiosperms: O-acetyl-4-O-methylglucuronoxylan, R: H or acetyl group, Ac: acetyl group. In red are represented the xylanolytic enzymes involved in its degradation. 1: endoxylanase, 2: acetylxylan esterase, 3: α-glucuronidase, and 4: β-xylosidase.

In the case of the most frequent glucuronoxylan from angiosperms (hardwoods), O-acetyl-4-O-methylglucuronoxylan, its complete hydrolysis needs the cooperative action of the endo-1,4-β-xylanase (endoxylanase), acetyl esterase, α-glucuronidase, and β-xylosidase (Fig. **4**).

The degradation of the most extended glucomannan from gymnosperms (softwoods), the O-acetyl-galactoglucomannan, involves endomannanases, α-galactosidases, acetylglucomannan esterases, β-mannosidases, and β-glucosidases (Fig. **5**).

Xylanases are produced mainly by fungi, although they can also be found in marine algae, protozoans, crustaceans, insects, snails, and seeds of land plants. The most interesting organisms producing xylanolytic enzymes are both mesophilic and thermophilic filamentous fungi, which secrete elevated levels of enzymes into the medium much higher than those found in yeasts and bacteria [19]. Due to the important biotechnological applications of fungal xylanases many of these enzymes have been isolated, characterized, and heterologously overexpressed. Commercial applications of xylanases demand thermostable enzymes showing optimal activity at alkaline pHs [20].

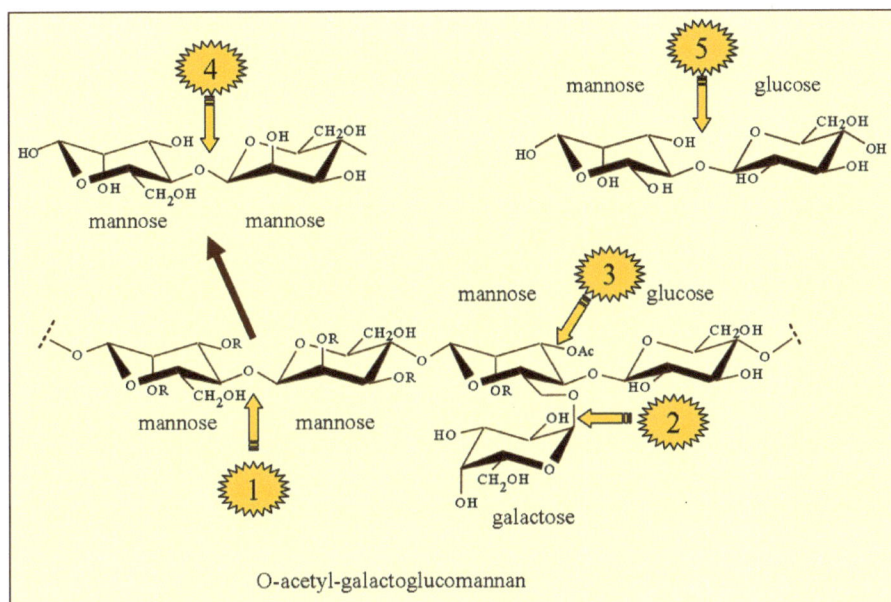

Figure 5: Structure of the most frequent hemicellulose in gymnosperms: O-acetyl-galactoglucomannan, R: H or acetyl group, Ac: acetyl group. In yellow are represented the enzymes involved in its degradation. 1: endomannanase, 2: α-galactosidase, 3: acetylglucomannan esterase, 4: β-mannosidase, and 5: β-glucosidase.

Among the mesophilic fungi, the genera *Aspergillus* and *Trichoderma* are the best studied xylanase producers. Regarding the thermophilic fungi, the better characterized enzymes are secreted by the species *Chaetomium thermophile*, *Humicola insolens*, *Humicola lanuginosa*, *Humicola grisea*, *Melanocarpus albomyces*, *Paecylomyces variotii*, *Talaromyces byssochlamydoides*, *Talaromyces emersonii*, *Thermomyces lanuginosus*, and *Thermoascus aurantiacus* [19]. As in mesophilic fungi, a multiplicity of xylanases differing in stability, catalytic efficiency, and activity on different substrates have been described in thermophilic fungi [16]. The optimal temperatures of these fungal xylanases diverge from 60 to 80 °C and the optimal pH of most of them ranges from 4.5–6.5. The diversity of xylanase isoenzymes of different molecular masses (from 3 to 38 kDa) might be related with their diffusion into the plant cell walls.

β-xylosidases are less common than xylanases. Most of them are cell-bound, and those produced by filamentous fungi are retained within the mycelium. Generally, they have relatively high MWs, between 60 and 360 kDa. Most of them have acidic pH optima (4.0-5.0) and the optimum temperature varies from 40 to 80 °C.

The accessory enzymes acetylxylan esterases or α-glucuronidase have been less studied although they are critical in the early steps of hemicellulose utilization. Acetylxylan esterase has been described in *T. reesei*, *Schizophylum commune*, and *Aspergillus niger* [18]. The filamentous fungi appear to be poor producers of α-glucuronidase, although the enzyme has been found in some of them such as *S. commune* [21], several species of *Trichoderma* and *Tyromyces palustris* [18].

Glucomannan-degrading enzymes are produced by a vast variety of bacteria, yeasts, and fungi. The microbial mannan-degrading enzymes are grouped into glycoside hydrolase families 5 and 26 [22]. The best known mannolytic group among fungi belongs to genera *Aspergillus*, *Agaricus*, *Ceriporiopsis*, *Humicola*, *Phanerochaete*, *Sclerotium*, or *Trichoderma* [22, 23]. Mannan-degrading enzymes from *T. reesei*, *Aspergillus acualetus*, or *Agaricus bisporus* have been characterized and classified into family 5. However, the mannan-degrading enzymes from *H. insolens* belong to family 26. Although the xylanolytic enzymes have been described far more frequently than mannanolytic enzymes, many different mannan-degrading enzymes have been characterized in the last decade due to their potential applications (see below). The physicochemical properties vary from one another. The mannanases optimum temperature ranges from 50 to 74 °C, while most of them show an optimum activity at acidic pH (3.0-4.5). Other mannan-degrading enzymes such as α-galactosidases, β-mannosidases or acetyl glucomannan esterases from *Aspergillus* have been reported [15, 24, 25].

OXIDATIVE BIODEGRADING SYSTEMS

Lignin Biodegradation

Due to the aromatic nature and the structural complexity of lignin, its insolubility, and its high molecular weight, this macromolecule resists to the attack by most microorganisms. In fact, lignin biodegradation is the key step in wood decay and in LCMs bioconversions to valuable products [5, 26].

In nature, only the wood-rotting basidiomycetes WRF are able not only to solubilize and depolymerize lignin, but also to mineralize it to CO_2. The best studied are *Phanerochaete chrysosporium*, *Pleorotus ostreatus*, *Bjerkandera adusta*, *Pycnosporus cinnabarinus*, *Ceriporiopsis subvermispora*, *Phlebia* sp., or *Trametes versicolor*.

Lignin biodegradation process is carried out mainly through non-specific oxidative reactions. The oxidants responsible for ligninolysis must hold the following requirements: i) they must attack the non-phenolic lignin structures (which comprise 80-90% in lignin) and ii) they must be small enough to penetrate the secondary walls because it has been demonstrated that the porosity in intact lignocelluloses is too low to allow enzymes penetration [27]. However, it is well known that ligninolytic microorganisms secrete different types of enzymes (usually named as "ligninolytic enzymes") that are associated to lignin biodegradation. The precise role of such ligninolytic enzymes is unclear because none of them is able to delignify intact lignocellulose *in vitro*. Furthermore, the fact that they are able to cleave lignin model compounds and depolymerize synthetic lignins points to the direction that they certainly contribute to ligninolysis. Apparently, these enzymes act using low molecular weight mediators to carry out the ligninolytic process.

The WRF produce four major types of oxidoreductases or lignin degrading enzymes: lignin peroxidase (LiP, EC 1.11.1.14), formerly named as ligninase, manganese dependent peroxidase or manganese peroxidase (MnP, EC 1.11.1.13), versatile peroxidase (VP, EC 1.11.1.16), and laccase (EC 1.10.3.2). The action of these main enzymes is enhanced by the cooperation of the other accessory enzymes that generate the H_2O_2 required by the peroxidases and dehydrogenases that reduce lignin-derived compounds. The most studied accessory enzymes are glyoxal oxidase (EC 1.2.3.5), aryl-alcohol oxidase (veratryl alcohol oxidase, EC 1.1.3.7), pyranose 2-oxidase (glucose 1-oxidase, EC 1.1.3.4), cellobiose/quinone oxidoreductase (EC 1.1.5.1), and cellobiose dehydrogenase (EC 1.1.99.18). The availability of the *P. chrysosporium* genome [28, 29] has allowed the finding of other proteins that could be involved in lignin degradation. For instance, four sequences with homology to multicopper oxidases (MCOs), which belong to the same family of laccases, have been found. The precise role of these MCOs remains unclear, but the authors suggest that one intriguing possibility is the modulation of Fenton reactions, through the Fe^{2+} oxidation [28].

Peroxidases

LiP, MnP, and VP are extracellular hydrogen-peroxide requiring hemo-proteins with catalytic cycles similar to horseradish peroxidase in many aspects [5, 7, 27].

LiP was the first discovered ligninolytic enzyme in the fungus *P. chysosporium* [8], but it has been found in many WRFs. In *P. chrysosporium* and in most of the WRFs LiP is present as a series of isoenzymes encoded by different genes. In *P. chrysosporium* 10 LiP isoenzymes were characterized [27]. The recently published genome has confirmed all the previous results [29]. The LiP isoenzymes are glycoproteins of 38-46 kDa and pI values ranging from 3.2 to 4.0 [7]. Its catalytic cycle and its oxidative potential have been studied using single-ring aromatic compounds and phenolic and non-phenolic lignin model dimers. All these studies have allowed to know that LiP is able to oxidize phenolic compounds, amines, aromatic ethers, and polycyclic aromatics. Hydroquinones and substituted phenols (typical substrates of plant peroxidases) are not oxidized by LiP. The catalytic mechanism includes the attraction of one electron from the aromatic rings producing an intermediate cation radical that spontaneously may react to form ring-cleavage products, Cα-Cβ cleavage, side chain cleavage, demethylation, intramolecular addition, or rearrangements (for a recent review see [7]). As we have mentioned above, since LiP is too large to enter the plant cell, an indirect oxidation by LiP of low-molecular-weight diffusible compounds capable of penetrating the cell wall has been suggested. The most probable candidate is veratryl alcohol (VA), which is a metabolite produced by *P. chrysosporium* at the same time that LiP. However, the role of VA as an intermediate in lignin degradation is unclear because although the cation radical $VA^{.+}$ produced by LiP can diffuse from the enzyme and acts as diffusible oxidant, there is not a consensus among the scientific community about if the half-life of such cation radicals is long enough to carry out this role [30, 31].

MnPs are molecularly very similar to LiPs although usually they have slightly higher molecular masses (ranging from 40 to 60 kDa) and acidic pIs [32]. They have been found in many WRF and they seem to be more widespread that LiPs [33]. They are also produced as a series of isoenzymes coded by different genes. MnP preferentially oxidizes Mn^{2+}, always presents in wood and soils, into highly reactive Mn^{3+}. This Mn^{3+} is dissociated from the enzyme and must be stabilized by organic acids chelators [34]. Mn^{3+} is able to oxidize a variety of phenolic compounds, including amines, dyes, and phenolic lignin substructures and dimers, but it cannot attack non-phenolic units of lignin. For non-phenolic substrates, the oxidation by Mn^{3+} involves the formation of highly reactive radicals in the presence of a mediator [7]. MnP also is able to oxidize non-phenolic lignin model compounds *via* peroxidation of lipids [35, 36].

VPs from different *Bjerkandera* and *Pleurotus* species have been characterized. These enzymes exhibit MnP-like activity: oxidation of Mn^{2+}; LiP-like activity: oxidation of phenolic and non-phenolic lignin structures, veratryl alcohol and p-dimetoxybenzene [32]. Moreover, it is also able to oxidize susbtituted phenols and hydroquinones like plant peroxidases [37].

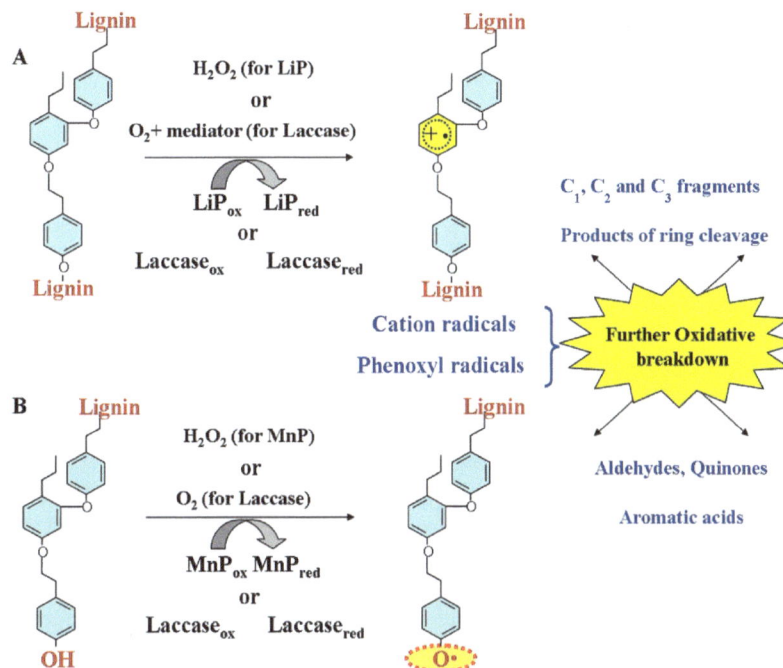

Figure 6: Extracellular oxidative degradation of non-phenolic arylglycerol β-aryl ether lignin (A) and terminal phenolic arylglycerol β-aryl ether lignin (B). VP can attack both structures (see text for details).

Laccases

Laccases are blue copper enzymes that belong to the multicopper oxidase family. They have been found in many Ascomycetes and Basidiomycetes, and they have been detected in most of the ligninolytic fungal species. In fact, more than 115 laccases have been purified and characterized in a wide variety of fungi. Their molecular masses range from 61 to 383 kDa, and their pI vary among 2.6 and 6.9, being the typical laccase a protein of 60-70 kDa with an acidic pI around 4.0. Like peroxidases, laccases are glycoproteins and many fungi produce different isoenzymes. These enzymes catalyze the oxidation of four molecules of the reducing substrate coupled to the four electron reduction of O_2 to H_2O. They are able to oxidize ortho-, meta-, or para- substitutes phenols, polyphenols, polyamines, aromatic amines, tiols and diarylamines [38, 39].

Laccase is also able to catalyze the oxidation of Mn ions in presence of chelators. Additionally, many low molecular weight compounds can be also oxidized by laccase and their radicals can in turn act as mediators to oxidize non phenolic compounds (revised in [38]). The most probable mediators in lignin degradation in nature are the so called natural mediators, which are derived from oxidized lignins and fungal metabolisms, such as phenolic aldehydes, ketones, acids, and esters related to the three lignin units including p-coumaric acid, vanillin, acetovanillone, methyl vanillate, syringaldehyde, and acetosyringone [40, 41].

In conclusion, the established simplified model is that the cooperative action of extracellular peroxidases, laccases, and other accessory enzymes, that are responsible for generating highly reactive and non-specific free radicals, which further undergo different reactions, catalyzes the lignin depolymerisation and degradation (Fig. **6**).

The broad substrate specificity of these ligninolytic peroxidases and laccases confers them an extraordinary potential in many biotechnological processes that will be discussed below.

BIOTECHNOLOGICAL AND INDUSTRIAL APPLICATIONS

Transformation of LCMs into Biofuels: Use of Cellulases and Hemicellulases

As we have previously mentioned, the increasing use of energy, the limits to the quantities of fossils resources, the accumulation of atmospheric CO_2, and the increasing environmental awareness are resulting in a growing demand for the search of alternative to petroleum-based fuels. Many progresses in genetics, biotechnology, chemistry, and engineering are leading to a new concept for converting biomass to valuable fuels and products, generally referred as biorefinery [42]. Biodiesel and bioethanol are the best alternative that will not only reduce greenhouse gas emissions, but also will offer great benefits in terms of security of supply. Biodiesel is produced from vegetable oils using chemical modifications. The bioethanol for the fuel market is currently obtained from easily fermentable agricultural materials such as sugar cane (in Brazil), sugar beet, corn starch, and diverse cereal grains (in Europe and USA). Generally, experts agree in that in a near future bioethanol and biodiesel will make an important contribution to our overall energy needs [10]. In fact, the US Department of Energy has proposed to replace the 30% of transportation fuels by biofuels and the European Commission is planning to substitute up to 25% of fossil fuels by alternative fuels by 2030 [43, 44]. Obviously, this future large-scale use of bioethanol will have to be based on production from LCMs because the agricultural feedstock will be not sufficient. LCMs are renewable, cheap, and abundantly available. However, there is much controversy whether the availability of lignocellulose is sufficient to meet the basic demand for material, animal feed, biobased chemicals, and fuels.

In the last decade much research is being made all over the world in order to improve biological conversion of LCMs to ethanol, with the main objective of bringing lignocellulosic ethanol towards industrial production [43].

The current production of ethanol from sugars or starch includes a pretreatment step to made substrate soluble, a hydrolysis step to obtain glucose from cellulose and a fermenting step to get ethanol from the sugars previously liberated. When lignocellulosic biomass is used as substrate it is necessary the use of pretreatments for separating lignin, hemicellulose, and cellulose, so these processes are much more complicated than just fermentation of C6 sugars. Moreover, pentoses have to be fermented together with hexoses. In consequence, these technologies are still far from being cost effective as compared to the production of bioethanol from sugar and starch (Fig. **7**).

Common pretreatment methods used to remove lignin and depolymerize cellulose and hemicellulose to soluble sugars, include physical treatments as ball milling, compression milling, cryomilling, etc. [4], or chemical treatments such as explosion of fibers using steam or ammonia, the treatment of biomass with diluted or concentrated acid, and hot water pretreatments [9, 26, 45]. In alkali-catalyzed pretreatments a part of lignin is removed and hemicelluloses have to be hydrolyzed by the use of hemicellulases. But many of these pretreatments solubilize the hemicelluloses fraction into the liquid phase and the solid phase comprises lignin and cellulose. The hemicellulose fractions, previously hydrolyzed to monosaccharides, can be used as substrate to obtain compounds such as organic acids.

Lignin can be exploited to obtain other value-added products as aromatic compounds or bio-oil [46]. Alternative biological pretreatments have been proposed either to replace the physicochemical treatment, or for detoxification or specific removal of inhibitors previous to the fermentation process. Most of the proposed biological treatments use the ability of the aforementioned WRF of degrading lignin. The main advantages of biological delignification are the mild reaction conditions and less energy demand [47]. Nevertheless, the negative aspects of such promising biological processes are the long treatment times and the partial degradation of carbohydrates [4]. Since the degradation of lignin is one of the most crucial steps in the transformation of LCMs into ethanol, a new approach based in the rational modification of lignin biosynthesis using new lignin engineering strategies, is being carried out [26, 48]. Feedstock crops with cell-wall structures more susceptible to pretreatment and thus, more prepared to hydrolysis, or that are sufficiently altered that they would not require pretreatment, are under research in different laboratories. This will make biofuels a process economically viable [11].

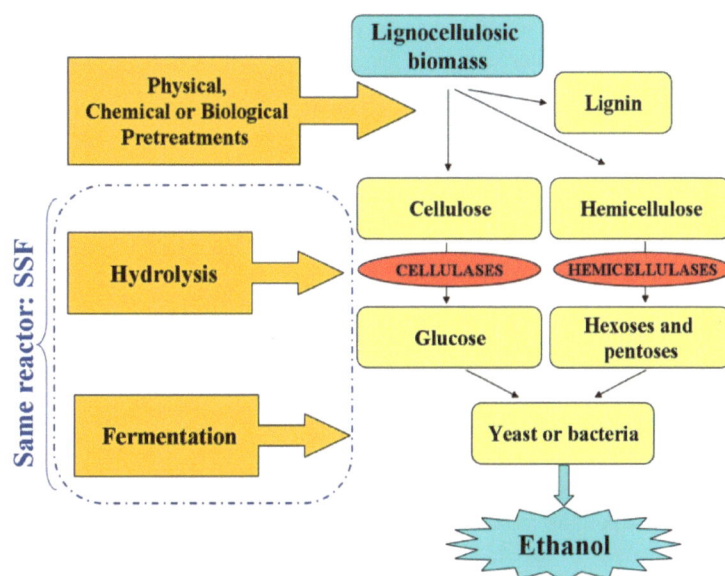

Figure 7: Simplified process for the bioconversion of LCMs into ethanol.

Finally, after the pretreatment, cellulose is subjected to acid or enzymatic hydrolysis (EH) with cellulases to get glucose. This monosaccharide is fermented by the yeast *Saccharomyces cerevisiae* or by the bacteria *Zymomonas mobilis* (Fig. **7**).

Another difference when LCMs are used for ethanol production is that, in addition of glucose, many other products are generated during the pretreatments of the raw material. For instance, hardwood or agricultural materials contain a considerable portion of pentoses such as xylose. Currently, the fermentation of a mixture of hexoses and pentoses is not efficient because no-wild organisms have been found that can ferment all sugars to ethanol. However, xylose can be fermented into ethanol by other microorganisms. Many research efforts are devoted to the development of improved strains of fermentative microorganisms capable of fermenting pentoses into ethanol, such as *Pichia stipitis* [43, 49]. An alterative to previously described two steps, separate hydrolysis and fermentation (SHF), is the so called simultaneous saccharification and fermentation (SSF), in which hydrolytic enzymes and fermentative microorganism(s) are present in the reactor [50]. Moreover, the optimal process integration can be achieved by transformation of lignocellulosic biomass in a single step that involves the production of cellulases and hemicellulases, the hydrolysis of carbohydrate polymers, and the fermentation of resulting sugars to desired products [17]. This process has been called consolidated bioprocessing (CBP).

Other bio-energy products that can be obtained from LCMs are bio-methane, hydrogen, or carbon dioxide [51, 52]. However, further advances in these technologies are needed in order to get cost effective technologies that can meet the challenges of large-scale utilization of LCMs.

Ligninolytic Fungi and Lignocellulolytic Enzymes in the Pulp and Paper Industries

Approximately 340 million tons of papers are produced world-wide every year. Pulp and paper industries are the largest user of renewable raw materials for consumer products. The most important concerns in the pulp and paper sector include bleaching with low-environmental impact, saving energy during pulping and bleaching, fiber recycling, and effluent minimization and detoxification [53]. In order to find solutions to the above challenges, the forest products industry has assayed the use of enzymes for many years. Ligninolytic fungi and lignocellulolytic enzymes are used in different steps of the pulp and paper-making. Pulping starts with the conversion of wood or agricultural materials into flexible fiber pulp that can be transformed into paper [54]. In this process, lignin must be separated and degraded. Several pulping processes are used, depending on the final applications, but the most frequent combine mechanical and chemical (especially kraft pulping, sulphite or alkaline cooking) technologies.

Biopulping is defined as the pretreatment of lignocellulosics raw material such as wood chips with lignin degrading fungi or enzymes before making the mechanical or chemical pulps treatment. These biotreatments oxidized lignin polymer, thus making easier the removal of lignin from polysaccharides. The most important advantages are the reductions in energy consumption, the increasing in the yield of fibers, and the improvement of paper strength properties. Additionally, biopulping helps to remove wood extractives, alleviate pitch problems, and reduce environmental impact in mechanical and chemical pulping and papermaking [53].

Many WRF have been assayed for pretreatments of mechanical or chemical pulping [54, 55]. Biomechanical pulping using such lignin-degrading fungi has proven to be engineering and economically feasible [55]. Some biopulping research using lignocellulolytic enzymes has also been reported [53]. In the recent years, concerns for the preservation of the forest have focused research on alternative fibrous supplies for papermaking. The most promising are the non-woody materials or agricultural residues. Biopulping of these alternative fibers with several WRF, such as *C. subvermispora* or *Pleurotus*, reduces the electrical power used in the refining stage [56].

The use of enzymes in pulping has also been assayed and it has been shown that the process is economically competitive, especially because they save in chemical cost. However, the use of enzymes is limited due to the restricted action of these proteins under the strict process conditions, such as high pH or elevated temperature. A mixture of cellulases and hemicellulases allowed better delignification of the pulps and saving in bleaching. Other enzymes, such as laccases, reduce energy requirements in mechanical pulping increasing the strength of the paper at the same time. Endoglucanases decrease the viscosity and increase the reactivity of several types of chemical pulps [54].

After cooking the crude pulp, the resulting fibers must be washed and bleached. Other manipulations such as sizing, color addition, etc., are carried out depending on the final paper products (residual lignin and derived compounds). The goal of bleaching is to whiten the pulps by altering or removing the coloured components. Traditional-bleaching sequences for chemical pulps were based on bleaching with chlorinated reagents. Nowadays, they are being replaced by elemental chlorine free (ECF) and totally chlorine free (TCF) sequences. These two environmental friendly bleaching technologies employ mainly oxygen based-bleaching (oxygen, alkaline hydrogen peroxide, and ozone). The main difference is that, while in ECF bleaching the use of chlorine chemicals (e.g. chlorine dioxide or hypochlorite) is authorized, the TCF bleaching does not use chloride at all.

Pulp bleaching using ligninolytic fungi or their enzymes is known as biobleaching. In this technology the use of ligninolytic enzymes or hemicellulases decreases the amount of bleaching chemicals required to get the desirable brightness of pulps. Although to laboratory scale, many ligninolytic fungi are effective in biobleaching [1]. The difficulties found in the control of fungal growth have limited the usage of fungi and most of the research has focused on the use of xylanases and other enzymes. Viikari *et al.* [57] demonstrated for the first time that the xylanases enhanced bleaching of pulp and saved up to 25% chlorine-containing chemicals. Many studies on xylanases and their use in pulp bleaching support the feasibility of their industrial application [58]. In fact, prebleaching of pulps with xylanases is being applied in many plants, where the use of these hemicellulases has increased the bleaching speed of ECF and TCF processes. Many commercial xylanases are available, mostly containing xylanases from *Trichoderma* species. The effect of xylanases in bleaching is indirect and seems to act removing reactants with the bleaching agents or acting as obstacles to the bleaching agents [54].

Many experiments with the ligninolytic peroxidases MnP and LiP have been carried out. However, the high costs of these enzymes and the very precise requirements for hydrogen peroxide (due to the intrinsic inactivation of these enzymes by hydrogen peroxide) and low-molecular weight cofactors (veratryl alcohol for LiP, and manganese and organic acids for MnP) have limited their use, facilitating the further lignin removal by hydrolyzing xylan chains [54].

The use of laccases in the form of laccase-mediator system (LMS) in pulp bleaching has attracted the attention of many laboratories around the world due to their efficiency in pulps delignification and bleaching. LMS react with the phenolic and non-phenolic fractions of lignin to delignify pulps [59]. Different LMSs can substantially reduce the demand of bleaching chemicals or allow bleaching to higher brightness [53]. Moreover, when LMSs are appropriately incorporated into an ECF or TCF-bleaching sequence, they are able to obtain fully bleached pulps. Moreover, LMS are able to substitute chlorine-containing reagents in manufacturing of non-woody pulps [60, 61]. However, unlike xylanases, the industrial use of laccases, has so far been blocked by the limited availability of laccases stable and active under industrial conditions, the high cost and potential toxicity of mediators released. A way to solve this problem is the use of mediators from natural resources as have been demonstrated recently [62]. The combined use of xylanases and LMS in bleaching has also been proposed [63].

Other process where the lignocellulolytic enzymes are being successfully applied is in deinking process when recycled papers are used as secondary fibers. Most of the waste papers are printed with copy or laser printed toner. These synthetic polymers do not disperse during conventional repulping hence they are very difficult to remove. Cellulases are very effective in facilitating the deinking process by decreasing the interaction of ink and toner particles with fibers, facilitating the flotation process and subsequent steps [64]. Cellulases and xylanases are also useful in the refining and drainage phases in the fiber recycling process [54]. Moreover, several ligninolytic fungi have been proven to be efficient in the control of pitch [65]. The term pitch is applied to wood extractives responsible for black deposits on bleached pulps that decrease their final quality and produce also important problems in the mill circuits during pulp and paper manufacture.

Industrial Effluents Color Removal and/or Detoxification

The environmental protection agencies are becoming more restrictive regarding water discharge from industrial effluents. Nowadays, the pulp and paper mills are doing many efforts in order to reduce the contaminants in all the stages, including pulping, bleaching, and papermaking. However, those stages release waste waters that impose coloration and toxicity problems to the receiving waters, causing sometimes environmental hazards [66]. Residual lignins and lignin derivatives (thio- and alkali-, or chloro-lignins) are the cause of this brownish dark coloration. Some industries, beside the classical primary and secondary treatments, use physical and chemical tertiary treatments to remove the recalcitrant color. However, these treatments are expensive and do not eliminate the totality of lignins. A feasible alternative is the use of ligninolytic fungi. Biological decolorization has been achieved using fungal mycelia, pellets, or immobilized cells or enzymes [67-70].

Other fields under research using these fungi or their enzymes are the biotreatment of industrial effluents containing lignin-like polymers, as is the cases of textile, olive-oil, alcohol distilleries, beer factories, or leather industries. The textile industry uses synthetic dyes very diverse in chemical composition. These dyes are designed to resist fading on exposure to sweat, light, chemicals, and microbial attacks. Because of that, dye-containing effluents are hard to decolorize by the conventional biological treatments. A plethora of WRF, producing different ligninolytic enzymes, has been proven to decolorize different textile dye effluents (for a review see [71]).

The olive oil industry, depending on the oil extraction process, generates water effluents or semisolid sludges. Both of them contain polymeric phenolic compounds (similar in structure to lignin) that give to these wastes a characteristic recalcitrant brownish black color. Several WRF, producing ligninolytic enzymes, are able to decolorize and detoxify these effluents [72-77]. Those decolorization/detoxification processes have been correlated to the production of MnP, LiP and laccases.

Recently, the marine-derived fungi, which also produce ligninolytic enzymes, are emerging as a potent alternative to terrestrial WRF in the decolorization of effluents [78].

Other Uses of the Fungal Lignocellulolytic Enzymes

Ligninolytic enzymes are extracellular, nonspecific, and are able to oxidize a broad variety of compounds. These properties confer the capacity of being applied to soil bioremediation and xenobiotic degradation. Several studies have shown that WRF are able to mineralize xenobiotics of diverse nature such as polycyclic aromatic hydrocarbons (PAH) [79]. *P. chrysosporium* is able to oxidize pyrene, anthracene, fluoride, and benzo [α] pyrene by LiP and MnP. Laccases from different fungi are also able to degrade PHAs, but in this case the enzyme must be aided by small mediators (for a review see [79]). Several WRF or their enzymes are able to degrade trichlorophenols, alkenes, herbicides, and many other recalcitrant compounds [36, 76]. Moreover, the alternative uses of efficient laccase mediators of natural origin together with laccase for PAH transformation and detoxification are giving very promise results [41, 80, 81].

Laccases are used to bleach textile, such as the denim finishing processes, or the degradation of indigo color [39]. Even, they have been commercialized for preparing cork stoppers for wine bottles, whereby the enzyme diminishes the characteristic cork taint and/or astringency that is frequently imparted to aged bottled wine [82].

More recently, new applications of laccase and LMSs natural mediators have arisen, such as enzymatic polimerization and polymer functionalization of cellulose or lignin fibers for obtaining new biomaterials, organic synthesis of fine chemicals and derivation of active compounds. The most promising ones are the enzymatic preparations of different polymeric polyphenols as non-toxic alternative to the formaldehyde-based chemical production [83-85]. Ligninolytic enzymes are very effective in the manufacturing of amalgamated material such as fiberboard. In this respect, laccase has been proven to activate the fiberbound lignin during composite manufacturing, resulting in boards without toxic synthetic adhesives [39]. Also, laccase has been demonstrated to be very useful to synthesize products of pharmaceutical importance, such as the production of new antibiotics or hormone derivatives, new anti-cancer drugs or antioxidants (for recent revisions on organic synthesis by laccases see [86] and [87]).

The natural origins of lignocellulolytic enzymes make possible their use in the food industry. In fact, they are applied to many agro-industrial processes in order to reduce cost, improve in food-nutritional quality and safety, and in the new products generation. In these industries laccases are used in wine and beer stabilization, fruit juice processing, sweet beet pectin gelation, and baking [88].

Cellulases and hemicellulases have also many commercial uses. Actually, hemicellulose and cellulose, together with pectinases, account for 20% of the world enzyme market. Besides the uses of hemicellulases commented in the previous sections, commercial fungal xylanases are used in animal feed, bread-making, juice and wine industries, manufacture of bread, food and drinks, or xylitol production, etc. Additionally, they can be used in the textile industry to process plant fibers [19].

Cellulases are used in the textile industry in the treatment of cotton fabrics for improving the properties of the materials. They became widely used for biostoning of denim garment treatment as an alternative to stone-washing processes in order to achieve an aged look [89].

Mannanases are used in different food industries such as instant coffee and beet sugar syrup processing, or the juice and wine clarification. Incorporation of mannanases into some animal diets, improves the nutritional value of poultry feeds, resulting in an improvement of weight gain and feed conversion efficiency (for recent reviews see [22] and [90]).

Finally, cellulases in combination with amylases or mannanases are commonly added to the detergents. Mannanases cleaves the β-1,4-linkage between mannose units in mannan- containing guar gums, which are commonly used as a thickener or stabiliser in many types of household and personal care products and foods. Stains containing these gums are generally difficult to remove because mannans show a strong adherence to cellulose fibres. The treatments with enzymes which are able to break down the gum polymer into smaller and water soluble carbohydrate fragments reduce the reappearing stain processes [22, 90].

CONCLUSIONS AND FUTURE PERSPECTIVES

In nature, the most efficient biodegraders of lignocellulosic materials are the filamentous fungi. They have potent enzymatic systems that are able to depolymerize the three major macromolecules: cellulose, hemicelluloses and lignin. The enzymes involved in those processes are well known. Many have been deeply characterized, cloned and heterologously expressed. The crystallographic structure of a lot of them has been solved and numerous studies of structure-function have been carried out in order to determine their catalytic mechanisms. The genome sequences of some lignocellulolytic fungi have been obtained and novel gene products probably involved in lignocellulose degradation are emerging. New protein engineering techniques such as directed molecular evolution are being applied in order to get more efficient enzymes. All this knowledge, beside with all the experience accumulated from a lot of years of using these enzymes in many biotechnological applications, has made of fungal lignocellulolytic enzymes one of the most powerful tools for green biotechnology and environmental-friendly technologies.

LIST OF ABBREVIATIONS

BRF   : brown rot fungi

CBH   : cellobiohydrolase

CBP   : consolidated bioprocessing

ECF   : elemental chlorine free

EG   : endoglucanase

EH   : enzymatic hydrolysis

LCM   : lignocellulosic material

LiP   : Lignin peroxidase

LMS   : laccase-mediator system

MCO   : multicopper oxidase

MnP   : manganese dependent peroxidase

MW   : molecular weight

PAH   : polyaromatic hydrocarbon

SHF   : separate hydrolysis and fermentation

SSF   : simultaneous saccharification and fermentation

TCF   : totally chlorine free

VA   : veratryl alcohol

VP   : versatile peroxidase

WRF   : White rot fungi

ACKNOWLEDGEMENTS

We would like to thank to Ministerio de Educación y Ciencia (grant BFU2006-00972/BMC) and Junta de Andalucía (grants P06-CVI-1377 and BIO318) for financial support. AM-M is a postdoctoral fellow from Plan Propio (Universidad de Granada).

REFERENCES

[1]   Pérez J, Muñoz-Dorado J, de la Rubia T, Martínez J. Biodegradation and biological treatments of cellulose, hemicellulose and lignin: an overview. Int Microbiol 2002; 5: 53-63.

[2]   Reddy N, Yang Y. Biofibers from agricultural byproducts for industrial applications. Trends Biotechnol 2005; 23: 22-7.

[3]   Zhang YHP, Lynd LR. Toward an aggregated understanding of enzymatic hydrolysis of cellulose: noncomplexed cellulose systems. Biotechnol Bioeng 2004; 88: 797-824.

[4]   Kumar R, Singh S, Singh OV. Bioconversion of lignocellulosic biomass: biochemical and molecular perspectives. J Ind Microbiol Biotechnol 2008; 35: 377-91.

[5]   Martinez AT, Speranza M, Ruiz-Duenas FJ, *et al.* Biodegradation of lignocellulosics: microbial, chemical, and enzymatic aspects of the fungal attack of lignin. Int Microbiol 2005; 8: 195-204.

[6]   Sánchez C. Lignocellulosic residues: Biodegradation and bioconversion by fungi. Biotechnol Adv 2009; 27: 185-94.

[7]   Wong DWS. Structure and action mechanism of ligninolytic enzymes. Appl Biochem Biotechnol 2008; 10.1007/s12010-008-8279-z

[8]   Tien M, Kirk TK. Lignin-degrading enzyme from the hymenomycete *Phanerochaete chrysosporium* Burds. Science 1983; 221: 661-3.

[9]   Mosier N, Wyman C, Dale B, *et al.* Features of promising technologies for pretreatment of lignocellulosic biomass. Bioresour Technol 2005; 96: 673-86.

[10]   Percival Zhang YH. Reviving the carbohydrate economy *via* multiproduct lignocellulose biorefineries. J Ind Microbiol Biotechnol 2008; 35: 367-75.

[11]   Li X, Weng JK, Chapple C. Improvement of biomass throught lignin modification. Plant J 2008; 54: 569-81.

[12]   Beguin P, Aubert JP. The biological degradation of cellulose. FEMS Microbiol Rev 1994; 13: 25-58.

[13]   Shoham Y, Lamed R, Bayer EA. The cellulosome concept as an efficient microbial strategy for the degradation of insoluble polysaccharides. Trends Microbiol 1999; 7:275–81.

[14]   Lynd LR, Weimer PJ, Willem H, van Zyl WH, Pretorius IS. Microbial cellulose utilization: Fundamentals and Biotechnology. Microbiol Mol Biol Rev 2002; 66: 506-77.

[15]   Kirk K, Cullen D. In: Young RA, Akhtar M, Eds. Environmental friendly technologies for pulp and paper industry. Willey: New York, 1998; pp. 273-307.

[16]   Maheshwari R, Bharadwaj G, Bhat MK. Thermophilic fungi: their physiology and enzymes. Microb Mol Biol Rev 2000; 64: 461-88.

[17]   Lynd LR, van Zyl WH, McBride JE, Laser M. Consolidated bioprocessing of cellulosic biomass: an update. Curr Opin Biotechnol 2005; 16: 577-83.

[18]   Jeffries TW. In: Ratledge C, Ed. Biochemistry of microbial degradation. Kluwer: Dordrecht. 1994; pp. 233-77.

[19]   Polizeli MLTM, Rizzatti ACS, Monti R, Terenzi HF, Jorge JA, Amorim DS. Xylanases from fungi: properties and industrial applications. Appl Microbiol Biotechnol 2005; 67: 577-91.

[20]   Kulkarni N, Shedye A, Rao M. Molecular and biotechnological aspects of xylanases. FEMS Microbiol Rev 1999; 23: 41-56.

[21]   Tenkanen M, Siika-aho M. An alpha-glucuronidase of *Schizophyllum commune* acting on polymeric xylan. J Biotechnol 2000; 78: 149-61.

[22]   Moreira LRS, Filho EXF. An overview of mannan structure and mannan-degrading enzyme systems. Appl Microbiol Biotechnol 2008; 79: 165-78.

[23]   Dhawan S, Kaur J. Microbial mannanases: an overview of production and applications. Crit Rev Biotechnol 2007; 27: 197-216.

[24]   Ademark P, Larsson M, Tjerneld F, Stålbrand H. Multiple α-galactosidases from *Aspergillus niger*: purification, characterization and substrate specificities. Enzyme Microb Technol 2001; 29: 441-8.

[25]   Athanasopoulos VI, Niranjan K, Rastall RA. The production, purification and characterisation of two novel α-D-mannosidases from *Aspergillus phoenicis*. Carbohydr Res 2005; 340: 609-17.

[26]   Weng JK, Li X, Bonawitz ND, Chapple C. Emerging strategies of lignin engineering and degradation for cellulosic biofuel production. Curr Opin Biotechnol 2008; 19: 166-72.

[27]   Hammel KE, Cullen D. Role of fungal peroxidases in biological ligninolysis. Curr Opin Plant Biol 2008; 11: 349-55.

[28]   Kersten P, Cullen D. Extracellular oxidative systems of the lignin-degrading Basidiomycete *Phanerochaete chysosporium*. Fungal Genet Biol 2007; 44: 77-87.

[29]   Martínez D, Larrondo LF, Putnam N, *et al.* Genome sequence of the lignocelullose degrading fungus *Phanerochaete chysosporium* strain RP78. Nat Biotechnol 2004; 22: 695-700.

[30]   Bietti M, Baciocchi E, Steenken S. Lifetime, reduction potential and base induced fragmention of the veratryl alcohol radical cation in aqueous solution. Pulse radiolysis studies on a ligninase mediator. J Phys Chem A 1998; 102: 7337-42.

[31]   Kapich AN, Jensen KA, Hammel KE. Peroxyl radicals are potential agents of lignin biodegradation. FEBS Lett 1999; 461: 115-9.

[32]   Martinez AT. Molecular biology and structure-function of lignin-degrading heme peroxidases. Enzyme Microb Technol 2002; 30: 425-44.

[33]   Hofrichter M. Review: lignin conversion by manganese peroxidase (MnP). Enzyme Microb Technol 2002; 30: 454-66.

[34]   Pérez J, Jeffries TW. Roles of manganese and organic acid chelators in regulating lignin biodegradation and biosynthesis of peroxidases by *Phanerochaete chrysosporium*. Appl Environ Microbiol 1992; 58: 2402-9

[35]   Reddy GVB, Sridhar M, Gold MH. Cleavage of nonphenolic β-1 diarylpropane lignin model dimers by manganese peroxidase from *Phanerochaete chrysosporium*. Eur J Biochem 2003; 270: 284-92.

[36]   Wariishi H, Valli K, Renganathan V, Gold MH. Thiol-mediated oxidation of nonphenolic lignin model compounds by manganese peroxidase of *Phanerochaete chrysosporium*. J Biol Chem 1989; 264: 14185-91.

[37]   Heinfling A, Martínez MJ, Martínez AT, Bergbauer M, Szewzyk U. Transformation of industrial dyes by manganese peroxidases from *Bjerkandera adusta* and *Pleurotus eryngii* in a manganese-independent reaction. Appl Environ Microbiol 1998; 64:2788-93

[38]   Badrian P. Fungal laccases-ocurrence and properties. FEMS Microbiol Rev 2006; 30: 215-42.

[39]   Rodríguez Couto S, Toca Herrera JL. Industrial and biotechnological applications of laccases: a review. Biotechnol Adv 2006; 24: 500-13.

[40]   Camarero S, Ibarra D, Martinez MJ, Martinez AT. Lignin-derived compounds as efficient laccase mediators for decolorization of different types of recalcitrant dyes. Appl Environ Microbiol 2005; 71: 1775-84.

[41]   Cañas AI, Alcalde M, Plou F, Martínez MJ, Martínez AT, Camarero S. Transformation of polycyclic aromatic hydrocarbons by laccase is strongly enhanced by phenolic compounds present in soil. Environ Sci Technol 2007; 41:2964-71.

[42]   Ragauskas AJ, Williams CF, Davison BH, *et al.* The path forward for biofuels and biomaterials. Science 2006; 311: 484-9.

[43]   Hahn-Hagerdal B, Galbe M, Gorwa-Grauslund MF, Liden G, Zacchi G. Bioethanol: the fuel of tomorrow from the residues of today. Trends Biotechnol 2006; 24: 549-56.

[44]   Himmel ME, Ding SY, Johnson DK, *et al.* Biomass recalcitrance: engineering plants and enzymes for biofuels production. Science 2007; 15: 804-7.

[45]   Hendriks ATWM, Zeeman G. Pretreatments to enhance the digestibility of lignocellulosic biomass. Bioresour Technol 100: 10-8.

[46]   Stöker M. Biofuels and biomass-to-liquid fuels in the biorefinery: catalytic conversion of lignocellulosic biomass using porous materials. Angew Chem Int Ed Engl 2008; 47: 9200-11.

[47]   Lee J. Biological conversion of lignocellulosic biomass to ethanol. J Biotechnol 1997; 56: 1-24.

[48]   Vanholme R, Morreel K, Ralph J, Boerja W. Lignin engineering. Curr Opin Plant Biol 2008; 11: 278-285.

[49]   Jeffries TW, Shi NQ. Genetic engineering for improved xylose fermentation by yeasts. Adv Biochem Engi/Biotechnol 1999; 65: 117–61.

[50]   Olofsson K, Bertilsson M, Liden G. A short review on SSF: an interesting process option for ethanol production from lignocellulosic feedstocks. Biotechnol Biofuels 2008; 1: 1-14.

[51]   Gunaseelan VN. Biochemical methane potential of fruits and vegetable solid waste feedstocks. Biomass Bioenergy 2004; 26: 389-99.

[52]   Yuan JS, Tiller KH, Al-Ahmad H, Stewart AR, Stewart CN. Plants to power: bioenergy to fuel the future. Trends Plan Sci 2008; 13: 421-9.

[53]   Widsten P, Kandelbaue, A. Laccase applications in the forest products industry: a review. Enzyme Microb Technol 2008; 42: 293-307.

[54]   Kenealy WR, Jeffries TW. In: American Chemical Society, Ed. Wood deterioration and preservation: advances in our changing world. Washington: DC, Oxford University Press 2003; pp. 210-39.

[55]   Akhtar M, Scott GM, Swaney RE, Kirk TK. In: Eriksson KEL, Cavaco-Paulo A, Eds. Enzyme applications in fiber processing. Washington, DC: American Chemical Society 1998; pp. 15-27.

[56]   Dorado J, Almendros G, Camarero S, Martinez AT, Vares T, Hatakka A. Transformation of wheat straw in the course of solid-state fermentation by four ligninolytic basidiomycetes. Enzyme Microb Technol 1999; 25: 605-12.

[57]   Viikari L, Ranua M, Katelinen A, Linko A, Sundquist L. Proceedings of the 3th international conference on biotechnology in pulp and paper industry. Stockholm, Sweden 1986; pp. 67-9.

[58]   Bajpai P. Biological bleaching of chemical pulps. Crit Rev Biotechnol 2004; 24: 1-58.

[59]   Call HP, Mücke I. History, overview and applications of mediated lignolytic systems, especially laccase-mediator-systems (Lignozym®-process). J Biotechnol 1997; 53: 161-202.

[60]   Camarero S, Garcia O, Vidal T, *et al.* Efficient bleaching of non-wood high-quality paper pulp using laccase-mediator system. Enzyme Microb Technol 2004; 35: 113-20.

[61]   Ibarra D, Camarero S, Romero J, Martínez MJ, Martínez AT. Integrating laccase-mediator treatment into an industrial-type sequence for totally chlorine-free bleaching of eucalypt kraft pulp. J Chem Technol Biotechnol 2006; 81:1159-65.

[62]   Camarero S, Ibarra D, Martinez AT, Romero J, Gutiérrez A, del Río JC. Paper pulp delignification using laccase and natural mediators. Enzyme Microb Technol 2007; 40: 1264-71.

[63]   Valls C, Roncero B. Using both xylanase and laccase enzymes for pulp bleaching. Bioresour Technol 2009; 100: 2032-9.

[64]   Heise OU, Unwin JP, Klungness JH, Fineran WG, Sykes M, Abubakr S. Industrial scale up of enzyme-enhanced deinking of nonimpact printed toners. Tappi J 1996; 9: 207-12.

[65]   Gutiérrez A, del Río JC, Martinez MJ, Martinez AT. The biotechnological control of pitch in paper pulp manufacturing. Trends Biotechnol 2001; 19: 340-8.

[66]   Ali M, Sreekrishnan TR. Aquatic toxicity from pulp and paper mill effluents: a review. Adv Environ Res 2001; 5: 175-96.

[67]   Garg SK, Modi DR. Decolorization of pulp-paper mill effluents by white-rot fungi. Crit Rev Biotechnol 1999; 19: 85-112.

[68]   Pérez J, Sáez L, de la Rubia T, Martínez J. *Phanerochaete flavido-alba* ligninolytic activities and decolorization of partially bio-depurated paper mill wastes. Water Res 1997; 31: 495-502.

[69]   Sahoo DK, Gupta R. Evaluation of ligninolytic microorganisms for efficient decolorization of a small pulp and paper mill effluent. Proc Bio 2005; 40: 1573-8.

[70]   Thompson G, Swain J, Kay M, Foster CF. The treatment of pulp and paper mill effluents: a review. Bioresour Technol 2001; 77: 275-86.

[71]   Wesenberg D, Kyrikides I, Agathos SN. White-rot fungi and their enzymes for the treatment of industrial dye effluents. Biotechnol Adv 2003; 22: 161-87.

[72]   D'Annibale A, Crestini C, Vinciguerra V, Giovannozi Sermannin G. The biodegradation of recalcitrant effluents from an olive mill by a white rot fungus. J Biotechnol 1998; 61: 209-18.

[73]   D'Annibale A, Stanzi S, Vinciguerra V, Di Mattia E, Giovannozi Sermannin G. Characterization of inmobilized laccase from *Lentinus edodes* laccase and its use in olive-oil mill waste-water treatment. Process Biochem 1999; 34: 697-706.

[74]   D'Annibale A, Stanzi S, Vinciguerra V, Giovannozi Sermannin G. Oxirane-immobilized *Lentinus edodes* laccase: stability and phenolic removal efficiency in olive mill waste-water. J Biotechnol 2000; 77: 265-73.

[75]   Sayadi S, Ellouz R. Screening of white rot fungi for the treatment of olive mill waste waters. J Chem Technol Biotechnol 1993; 57: 141-6.

[76]   Pérez J, de la Rubia T, Ben Hamman O, Martínez J. *Phanerochaete flavido-alba* laccase induction and modification of manganese peroxidase isoenzyme pattern in decolorized olive oil mill wastewaters. Appl Environ Microb 1998; 64: 2726-9.

[77]   Berrio J, Plou FJ, Martinez AT, Martinez MJ. Immobilization of *Pynocporus coccineus* laccase on Eupergit C: stabilization and treatment of olive oil mill wastewaters. Biocatal Biotransform 2007; 25: 130-4.

[78]   Raghukumar C, D'Souza-Ticlo D, Verma AK. Treatment of colored effluents with lignin-degrading enzymes: an emerging role of marine-derived fungi. Crit Rev Microbiol 2008; 34: 189-206.

[79]   Peng RH, Xiong AS, Xue Y, *et al.* Microbial biodegradation of polyaromatic hydrocarbons. FEMS Microbiol Rev 2008; 32: 927-55.

[80]   Camarero S, Cañas AI, Nousiainen P, Record E, Lomascolo A, Martínez MJ, Martínez AT. P-hydroxycinnamic acids as natural mediators for laccase oxidation of recalcitrant compounds. Environ Sci Technol. 2008; 42: 6703-9.

[81]   Mayer AM, Staples RC. Laccase: new function to and old enzyme. Phytochem 2002; 60: 551-65.

[82]   Riva S. Laccases: blue enzymes for green chemistry. Trends Biotechnol 2006; 24: 219-26.

[83]   Fackler K, Kuncinger T, Ters T, Srebotnik E. Laccase-catalyzed functionalization with 4-hydroxy-3-methoxybenzylurea significantly improves internal bond of particle boards. Holzforschung 2008; 62: 223-9.

[84]   Kudanga, T, Prasetyo EN, Sipilä, J, Nousiainen P, Widsten P, Kandelbauer A, Nyanhongo1 GS, Guebitz G. Laccase-Mediated Wood Surface Functionalizatio, Eng Life Sci 2008, 8: 297-302.

[85]   Aracri E, Colom JF, Vidal T. Application of laccase-natural mediator systems to sisal pulp: an effective approach to biobleaching or functionalizing pulp fibres? Bioresource Technol 2009; 100: 5911-6.

[86]   Kunamneni A, Camarero S, García-Burgos C, Plou FJ, Ballesteros A, Alcalde M. Engineering and applications of fungal laccases for organic synthesis. Microb Cell Fact 2008; 20: 7-32.

[87]   Mikolasch, A, Schauer F. Fungal laccases as tools for the synthesis of new hybrid molecules and biomaterials. Appl Microbiol Biotechnol 2009; 82: 605-24

[88]   Minussi RC, Pastore GM, Duran N. Potential application of laccase in the food industry. Trends Sci Technol 2002; 13: 205-16.

[89]   Belghith H, Ellouz-Chaabouni H, Gargouri A. Biostoning of denims by *Penicillium occitanis* (Pol6) cellulases. J Biotechnol 2001; 89: 257-62.

[90]   Dhawan S, Kaur J. Microbial mannanases: an overview of production and applications. Crit Rev Biotechnol 2007; 2: 197-216.

CHAPTER 3

Genomic and Proteomic Analyses Provide Insights into the Potential of Filamentous Fungi for Biomass Degradation

Jean Marie François[1,2,3*] and Olivier Guais[1, §]

[1]Université de Toulouse; INSA, UPS, INP, 135 Avenue de Rangueil, F-31077 Toulouse; [2]INRA, UMR792 Ingénierie des Systèmes Biologiques et des Procédés, F-31400 Toulouse, [3]CNRS, UMR5504, F-31400 Toulouse, France and [§]Present address : Cinabio-Adisseo France S.A.S, 135 Avenue de Rangueil, 31077 Toulouse, France.

Abstract: Fungi are metabolically versatile organisms in Nature, existing either as free-living species, in association with other species, *e.g.* lichens, mycorrhiza, or as pathogens in animals and plants. They are characterized by their notable ability to degrade a wide variety of complex polysaccharides and recalcitrant waste organic materials. Their attractiveness in Biotechnology comes from the remarkable capacity to secrete a wide spectrum of enzymes that are used in food and for biomass degradation, as well as to produce a variety of secondary metabolites ranging from human therapeutics (*e.g.* antibacterial and antifungal agents) to specialty chemicals such as polyketides and organic acids. This minireview is mainly focused on filamentous fungi as a mycofactory for enzyme production dedicated for plant biomass (lignocellulose, hemicelluloses) degradation. As several filamentous fungi relevant to this industrial application have been recently sequenced, we will first provide an overview on the main fungal genome sequences, pointing out that 7 to 12 % of the gene content of these genomes codes for secreted proteins, collectively termed as the 'secretome', among which a significant proportion encodes putative 'Carbohydrate-Active enzymes'. In the second part, we will review how the exoproteome, which represents the set of secreted proteins in the medium, has been initially characterized, and then show that a combination of computational, transcriptomic and proteomics methods is the most effective approach to reveal the reliability of predicted secretomes. At this stage, two major observations can be made. Firstly, the exoproteome of a fungus is strongly tied to its culture conditions and/or nutrient source. Secondly, and more interestingly, a catalog of genes encoding putative Carbohydrate Active enzymes greater than ever expected has been revealed from genome sequencing. Moreover, transcriptome analyses of filamentous fungi cultivated under cellulolytic/hemicellulolytic conditions have shed light on an impressive collection of upregulated genes encoding putative secreted proteins with yet uncharacterized function. Altogether, these new findings show that there is still a long way to go for a comprehensive understanding of the fungal secretome, which is the basis for a rational development of optimized strains in white biotechnology.

INTRODUCTION

There are currently more than 60 fungal genome sequences publicly available (see http://fungalgenomes.org/wiki/Fungal_Genome_Links), among which are 12 plant pathogen fungi and two oomycetes [36]. A significant large number of fungal sequencing projects are in progress (see www.broadinstitute.org/science/projects/fungal-genome/).

In addition, the dramatic fall in the price of DNA sequencing, coupled with the next generation of sequencing technologies are influential arguments for sequencing many other exotic fungal species. This will increase even faster the massive data resources already available on gene inventory, ecological /pathogenic adaptation, allelic variability, evolution, etc.

For the purpose of this minireview, we have focused our review on genome-sequenced filamentous fungi of direct biotechnological relevance in food and renewable Energy. As indicated in Table **1**, most of them belong to the Ascomycota phylum, except for *Phanerochaete chrysosporium,* which is a member of the Basidiomycota.

This list is completed with three additional genomes from phytopathogenic fungi since, these fungi also contain relevant genomic information with respect to industrial biotechnology applications, besides the presence of a set of

***Address correspondence to Jean Marie François:** INRA, UMR792 Ingénierie des Systèmes Biologiques et des Procédés, F-31400 Toulouse, [3]CNRS, UMR5504, F-31400 Toulouse, France ; E-mail: fran_jm@insa-toulouse.fr

gene families that are likely specific for their pathogenic action in plants [35]. *Saccharomyces cerevisiae* [13] has been included in the table to highlight significant difference in the genome of this model yeast with that from the filamentous fungi.

ENZYMATIC POTENTIAL OF FILAMENTOUS FUNGI FOR BIOMASS DEGRADATION AS REVEALED BY GENOME SEQUENCING

Overview of the Fungal Genome

The genome size of filamentous fungi sequenced so far is in the range of 33 ± 4 Mbp, *i.e.* about 3 times greater than that of *S. cerevisiae*, with a % GC that is characteristic of mesophylic species able to grow between 15 and 45 °C (Table 1).

Table 1: General Feature of Filamentous Fungi Genomes.

Species	Genome size (Mb)	% GC (overall)	Protein coding genes	Average exon size (bp)	Average intron size (bp)	Av number intron per genes	Ref.
Aspergillus niger CBS513.88	33.9	50.4	11,200	370	97	2.57	[29]
Aspergillus nidulans FGSCA4	30.1	50.3	10,701*	429	63	1.8	[12]
Aspergillus fumigatus Af293	29.4	49.9	9,923*	516	112	1.8	[25]
Aspergillus oryzae RIB40	37	48	12,074	397	nd	nd	[20]
Magnaporthe grisea70-15	37.8	51.6	12,841	247	94	1.89	[8]
Fusarium graminearum F4	36.1	48.3	11,640	242	56	2.22	[7]
Neurospora crassa ORT74A	38.7	50	10,620	nd	134	1.7	[11]
Phanerochaete chrysosporium RP78	29.9	57	10,048	233	65	2.6	[23, 42]
Trichoderma reesei (sn hypocrea jecorina)RUTC30	33.9	52	9,129	508	120	3.1	[21]
Penicillium chrysogenum Wisconsin54-1255	32.2	48.9	12,943	434	87	2.2	[40]
Saccharomyces cerevisiae	12.2	38	6,607	-	§	-	www.stanford.com

*Updated from CADRE (http://www.cadre-genomes.org.uk/aspergillus_links.html [29]).

§ In yeast, only 283 introns of 6000 genes contains an intron with size between 100 and 400 bp.

The fungal genomes are predicted to contain around 10,000 genes encoding proteins (size equal or greater than 100 amino acids), with *Trichoderma reesei* containing the lowest gene number, and *P. chrysogenum*, the highest, even though both species have the same genome size. This amount of predicted genes is around 1.6 fold higher than that of the *S. cerevisiae* genome. A notable difference also with the yeast genome is the presence of introns in almost all filamentous fungal genes at a rate of about 2 introns per gene. Dedicated bioinformatic tools have been used to search for genes encoding secreted proteins, such as SignalP 3.0 [3], which detects the presence of a signal peptide in the N-terminus of the predicted protein, or TargetP [9], which can discriminate between proteins destined for plasma membrane, mitochondria or chloroplast. These *in silico* analyses allowed identification of 7 to 12% of the genes from filamentous fungi genomes to encode secreted proteins that is collectively termed the 'secretome, whereas the size of the yeast 'secretome' is limited to 3% of the genes from its genome (Table **2**). Although these tools have been trained on secreted proteins from this model yeast, such that the predicted secretome size may not be fully accurate, it clearly illustrates the remarkable secretory capacity of the filamentous fungi, in accordance with their unique capacity to digest their food components extracellularly through a process that involves secretion of hydrolytic and/or oxidative enzymes before absorption of the nutrients. However, the analysis of genomes has so far provided very few mechanistic insights into this extraordinary capacity of filamentous fungi to secrete proteins, coupled to the fact that membrane trafficking and vesicle budding are more diverse in *T. reesei* [21] and that the *A. niger* secretory system shares some components with mammalian system that is not present in *S. cerevisiae* [29]. In addition, massively parallel sequencing of the genome from two high-cellulase secreting strains of *T. reesei* (NG14 and its improved descendant RUT C30) using the Illumina Solexa technology was compared with the published genome sequence of *T reesei* QM6a from which they both derived. Apart from the fact that this study revealed a surprisingly high number of mutagenic events randomly distributed in several functional categories, including transport, transcription, secretion/ vacuole targeting and metabolism, it also pointed out the potential, though unexpected, implication of vacuoles in protein secretion in *T. reesei* [18]. Future research should therefore carefully consider this intracellular compartment as a target for improving protein production.

Table 2: Overview of the 'Secretome' and Carbohydrate Actives Enzymes from Sequenced Filamentous Fungi.

Species	Total secreted protein*	Glycosyl hydrolase#	Glycosyl transferase	Carbo-hydrate binding module	Carbo-hydrate esterase	Polysac-charide lyase	Ref.
Aspergillus niger CBS513.88	681/881	243	110	40	24	8	[39]
Aspergillus nidulans	981/894	247	91	36	29	19	[12]
Aspergillus fumigatus	nd	263	103	55	29	13	CAZY[§]
Aspergillus oryzae	nd	285	114	30	26	21	CAZY[§]
Magnaporthe grisea	1491	231	94	58	47	4	[8]
Fusarium graminearum	1442	243	110	61	42	20	[7]
Neurospora crassa	891	171	76	39	21	3	[11]
Phanerochaete chrysosporium RP78	769	200	68	45	19	4	[23, 42]
Trichoderma reesei (sn hypocrea jecorina)	663	200	103	36	16	3	[21]
Penicillium chrysogenum	980	214	100	49	22	9	[40]
Postia placenta	-	144	75	6	10	6	[22]
Saccharomyces cerevisiae	163	45	67	12	3	0	[44]

*Genes predicted to encode secreted proteins by SignalP (3), nd: not determined;. [§]CAZy at www.cazy.com.

Secretome Potential as Evaluated from Genome Sequence

The Carbohydrate-Active enzymes (CAZymes) database developed by Henrissat and coworkers (see [5] for the last updated version) is a useful tool for both classification and compilation of the glycosyl hydrolase, glycosyl transferase, lyase and esterases implicated in the degradation of complex polysaccharides. In quantitative terms, the fungal CAZOmes from the Ascomycota, whether it concerns free-living or plant- pathogen fungi, is roughly equivalent, encompassing about 400 genes, with a similar distribution of the proteins in the different subfamilies (Table **2**). Additional featuresfrom this table need further comments. Firstly, the size of glycosyl hydrolase is two to three times bigger than that of glycosyltransferase, whereas the opposite is true for other eukaryotic genomes, including *S. cerevisiae* (Table **2**), *Arabidopsis thaliana, Drosophila, Caenorhabitis elegans, Homo sapiens,* etc. This ratio is consistent with the fact that these latter organisms are more concerned with constructing rather than breaking down polysaccharides and glycoproteins. Secondly, several CAZymes involved in plant polysaccharide depolymerisation bear a carbohydrate - binding module (CBM). This module is defined as a contiguous amino acid sequence with a discrete fold bearing carbohydrate binding activity [34]. To date, these binding modules have been classified into 58 different families based on amino acid sequence, binding specificity or structure (see CAZY at http://www.cazy.org for further details).

In filamentous fungi, the main representatives CBMs are those for cellulose-binding (CBM1 and CBM13) and for chitin-binding (CBM14 and CBM18). The plant pathogen *M. grisea* and *F. graminearum* are the two fungi harboring the highest number of these CBM (80% of the total CBM reported in Table **2**), and especially, in chitin-binding domain.

The largest number of CBMs in these phytopathogenic fungi may be due to the fact that they have to remain bound to their substrates, such as hemicelluloses, pectin and cutin, while degrading these cell wall components to finally penetrate inside the plant cells. On the other hand, cellulose-binding modules are found in many hydrolytic enzymes of filamentous fungi involved in biomass (cellulose, hemicellulose) degradation, consistent with the fact that these CBMs improve the catalytic efficiency of these enzymes on insoluble substrates (for a review see [34]). A final remark is related to the polysaccharide lyase family, which is mainly implicated in the biodegradation of pectin. The genome of *Aspergilli* species is particularly rich in genes encoding this family of enzymes, whereas other filamentous fungi including *T. reesei* and *M. grisea* show a relative paucity of this protein family, with the exception of the plant pathogen *F. graminearum*. This difference actually reflects the variety of nutritional sources these filamentous fungi can use.

Enzymatic Potential for Hemicellulose and Cellulose Biodegradation

Major genes family encoding carbohydrates enzymes implicated in cellulose and hemicellulose degradation are summarized in Table **3**. Unexpectedly, *Trichoderma reesei*, which is the workhorse fungus for cellulase production, displays the lowest content of genes related to plant biomass degradation, whereas the plant pathogen *M. grisea* shows the largest gene content for this activity.

Table 3: A Survey of Genes Encoding Carbohydrate Actives Enzymes Implicated in Hemicellulase and Cellulases Degradation*

Fungal Species	GH3	GH5	GH6+GH7	GH10	GH11	GH12	GH43	GH51	GH54	GH62	GH61	GH74	Total
Aspergillus niger CBS513.88	16	8	4	1	4	3	10	3	1	1	7	1	51
Aspergillus nidulans FGSCA4	20	17	5	3	2	1	15	3	1	2	9	2	78
Aspergillus fumigatus Af293	6	1	5	4	3	4	18	2	0	2	7	2	50
Aspergillus oryzae RIB40	23	13	4	4	4	4	20	3	1	2	8	0	84
Magnaporthe grisea 70-15	19	13	9	5	5	3	19	3	1	3	17	1	98
Fusarium graminearumF4	0	2	2	5	3	0	16	2	1	1	13	1	49
Neurospora crassa ORT7A	9	7	8	4	2	1	7	1	1	0	14	1	55
Phanerochaete chrysosporium RP78	13	18	10	6	1	11	4	2	0	0	14	4	74
Trichoderma reesei (sn hypocrea jecorina)RUTC-30	3	2	2	1	4	1	2	0	2	1	3	1	22
Penicillium chrysogenum Wisconsin	17	13	3	10	1	3	14	3	1	1	14	0	80

*Listing of the major glycosyl hydrolases acting for hemicellulase and cellulose degradation. CAZY categories reported correspond to β-glucosidase/cellobiase (GH3); endoglucanase (GH5); cellobiohydrolase (GH6+GH7); endo 1,4-β-xylanase/ endo 1,3-β-xylanase (GH10); xylanases (GH11); β-xyloxidase/β-glycosidase (GH43); α-L-arabinofuranosidase (GH51, GH54, GH62) and endoglucanase (GH61) and xyloglucanase/ endoglucanase (GH74); nc : not checked

Gene redundancy is another characteristic of these families, ranging from 2 to 20 members in a same category. This redundancy for genes encoding hydrolytic enzymes suggests that the filamentous fungi can optimize their enzymatic cocktails according to growth optimization and environmental conditions (*e.g.* nature of the carbon source, pH, ionic strength, etc) and that similar but not identical enzyme functions are necessary to effectively breakdown a range of complex carbohydrates polymers, whose structure, physical state and accessibility vary widely upon the botanical context and the extent of decay. It is also interesting to notice that β-glycosidase/ cellobiase (GH3), endoglucanase (GH5) and exocellobiohydrolase (GH6+GH7) are particularly well represented in most saprophytic fungi able to develop on decaying wood such as in *P. chrysosporium* and *P. chrysogenum*. This is in accordance with their remarkable capability to degrade plant cell walls, whereas these gene families are less abundant in fungal species like *F. graminearum*, since it is a plant pathogen that invades plant cereals using mainly cutinase, pectate lyase and polygalacturonase to degrade cutin and pectin that coat all outer plant cell surfaces.

Very recently, the genome of the brown-rod *Postia placenta* has been published [22]. This fungus is a common inhabitant of forest ecosystems and is largely responsible for the destructive decay of wooden structures by a rapid depolymerisation of cellulose involving oxidative systems, instead of hydrolysis, as done by other fungi including the white-rod *P. chrysosporium*. This oxidative cleavage mechanism for cellulose decomposition by these wood-rotting basidiomycetes, formulated 40 years ago, has now received genomic support from the genome sequencing of *P. placenta* [22]. The genome sequence of this fungus is about 3 times larger than that of other fungi. This is due to the fact that the DNA used was from a dikaryon and not from a haploid strain. More importantly, this genome contains two times fewer genes encoding Carbohydrate-Active enzymes, and it is notably deficient in conventional cellulase-encoding genes (Table **2**). In contrast, this fungal genome is enriched in many genes supporting Fenton chemistry through the generation of extracellular H_2O_2, including copper radical oxidases, numerous iron reductase and permeases and quinone reductases. These genomic data support the existence of oxidative systems for cellulose degradation that operate by redox cycling of small-molecular mass quinines or other redox compounds leading to the production of free hydroxyl radicals. These hydroxyl radicals can in turn depolymerize cellulose by oxidizing chains ends ([2] for a review). Since this depolymerized material is known to be less amenable to cellulase action [30], this would explain the lack of genes encoding cellulases and hemicellulases in this filamentous fungus.

Besides cellulolytic and hemicellulolytic systems, much additional information can be extracted from genomic sequences to account for the potential use of filamentous fungi in biomass bioconversion. The genome of the *Aspergilli* species harbors a large set of genes for degradation of other carbohydrates, including starch and glycogen (α-amylases), inulin (inulinase), α and β-glucan (β-1,3, 1,4 glucanases), pectin (pectate lyase, feruloyl esterases, pectin methyltransferases, acetyl xylan esterase, etc), as well as genes encoding proteins involved in proteolytic degradation [20, 29]. The secretome of the white-rot basidiomycete *Phanerochaete chrysosporium* also contains a catalogue of genes encoding several potential lignin and manganese peroxidases, copper radical oxidases, extracellular FAD-dependent oxidases, and cytochromes P450. This gene repertoire is consistent with the fact that, unlike many other fungi, *P. chrysosporium* is highly efficient for depolymerization and mineralization of lignin, the second most abundant polymer on earth which is composed of amorphous and insoluble aromatic material [19].

CHARACTERIZATION AND EXPRESSION OF THE FUNGAL "EXOPROTEOME" FOR BIOMASS DEGRADATION

Definition of Secretome, Exoproteome and Limitation of their Analysis by Proteomics

As already stated above, the saprotrophic nature of several filamentous fungi and the capacity to grow on complex substrates, such as on decaying wooden structure, is mainly attributed to the secretion into the culture medium of a broad array of hydrolytic enzymes. These extracellular proteins represent only part of the 'secretome', which more largely describes the repertoire of proteins that are processed through the endoplasmic reticulum secretory pathway [14, 38]. The potential size of a fungal secretome can be estimated by computational approaches, which rely on the existence of signal peptide at 5'termini of the corresponding gene sequence [3] or on a combination of transmembrane domains and signal peptide in the proteins [17]. However, these bioinformatic tools leave aside proteins that can be exported at the cell surface in the absence of any recognizable signal [26]. In addition, this composition of 'secretome' encompasses proteins secreted in all cellular compartments (*i.e.* vacuole, mitochondria) and those secreted into the medium. The term 'extracellular proteome' or 'exoproteome' is therefore used to define the set of proteins that is secreted into the culture medium. The exoproteomes of any filamentous fungus that are produced according to its nutrient sources and culture conditions can now be analyzed using advanced proteomic instrumentation such as matrix assisted laser desorption ionization time-of-flight mass spectrometry (MALDI-TOF MS) and capillary liquid chromatography-nanoelectrospray ionization tandem MS (LC-ESI-MS/MS). However, the identification of the secreted proteins strongly depends on the accuracy of the genome annotation, which in the case of filamentous fungi is still far from optimal. In addition, large scale proteomics methods have serious limitations. Low- abundance, low molecular-weight and unstable proteins will be less likely detected. Highly glycosylated proteins or those adhering strongly to cell wall or retained to substrates could escape detection. Also, exhaustive protein identification cannot be limited to a single separation method as is often made using the two-dimensional electrophoresis (2-DE). Combination of several independent separation methods, as for instance 2-DE, 1 DE, and shotgun analyses, is advised for a maximal coverage of protein identification, mainly when the genome is not available, as this has been done recently to identify the secreted proteins present in a commercial enzyme cocktail (termed Rovabio™Exl) produced by the industrial soil deuteromyces *Penicillium funiculosum* [15].

The Exoproteome Reflects the Saprophytic or Pathogenic Lifestyle of the Filamentous Fungi

A survey of the literature published up to 2009 on filamentous fungal exoproteomes allows categorizing broadly these studies from three different perspectives. The first perspective was to discover extracellular proteins that were expressed by the filamentous fungi that accommodate with a saprophytic or pathogenic lifestyle or in accordance to the technological process. Most of these earlier studies were carried out while the genomes of these fungi were either partially or not sequenced at all. Therefore, the identification of proteins was done by comparison with homologue proteins in databases. The first contribution was a comparative analysis of the exoproteome from *Aspergillus flavus* cultivated on glucose, potatoes dextrose and rutin (quercetin 3-0-glycoside). A total of 51 unique proteins could be identified, of which one third was specific to rutin. In another comparative study in the white-rot *Pleurotus sapidus*, Zorn *et al.* [45] showed significant differences in the exoproteome of this fungus in submerged cultures on peanut and with that from cultures on glass wool. Suarez *et al.* [37] also reported significant qualitative differences of the exoproteome of *T. harzinium* cultivated on pure chitin versus that from culture on cells walls. The characterization of the industrial enzymatic cocktail Rovabio™Excel produced by *Pencillium funiculosum* 8/403 was also recently reported, revealing the presence of more than 50 proteins, among which were several glycosylhydrolytic, hemicellulolytic and proteolytic enzymes [15]. Altogether, these early works showed the power of the systematic proteomic analysis to gauge the adaptation of the fungus to its environment, although this analysis was limited by the lack of genome sequences, leading to a list of peptides with no homologue in protein databases.

The genome sequence of the fungus is therefore indispensable for a better characterization of the exoproteome. The availability of annotated genome with mass spectrometric techniques was used in a comparative analysis of extracellular proteins of *P. chrysosporium* grown under ligninolytic conditions and on softwood chips [31]. This has permitted the identification of exactly 25 proteins from a total of 37 spots from both conditions on 2-D gels stained by Coomassie blue. This study showed that under ligninolytic condition, the eight prominent spots were identified as lignin and manganese peroxidase, whereas on softwood, several glycoside hydrolases were detected in a relatively large abundance, consistent with the fact that this substrate is rich in cellulose and hemicelluloses (see Table **4**). A similar approach was carried out for the characterization of the exoproteome of *P. chrysosporium* grown on oak [1]. Apart from a technical problem that required specific pre-washing of the wood chips to reduce brown material that hampered the 2D-separation, this work led to the identification of 16 proteins, most of which corresponded to cellulolytic and hemicellulolytic enzymes (cellobiohydrolase I and II, cellobiose dehydrogenase, β-glycosidases, see Table **4**), whereas no lignin peroxidase, manganese peroxidase or laccases were detected. An even more exhaustive comparative proteomic analysis, supported by the availability of the genome sequence was carried out in *F. graminearum* cultivated *in vitro* under 13 different media or '*in planta*' during infection of wheat heads [28]. A total of 289 proteins (229 *in vitro* and 120 *in planta*) were identified. The high annotation quality of the *F. graminearum* genome allowed these authors to find that > 90 % of the *in vitro* proteins had a predicted signal peptide, whereas less than 50% of the secreted fungal proteins were found *in planta*. This result was explained mainly by a partial lysis of the fungus during its invasion in the plant [28]. Furthermore, and because of a reliable genome annotation, this study showed for the first time that proteins can be detected both in non-fractionated preparations directly after extraction from the culture broth or in extracts of plants after infection. This extensive analysis also revealed that the exoproteome was a true signature of the growth conditions of *F. graminerum*. A proteomic analysis of extracellular proteins was also performed to compare the protein profile between two hypersecretory *Trichoderma reesei* strains, the industrial strain CL847 and the reference overproducer strain RUT-C30, cultivated on lactose as the carbon source [16], and using the annotated genome sequence of *T. reesei* QM6A for protein assignment. As expected, the most intense spots identified in this proteomic analysis corresponded to cellobiohydrolase Cel7A and Cel6A, in agreement with the fact that these two proteins represent 70 -80% of the total proteins in the enzymatic cocktail produced by these industrial strains [24].

Table 4: Identification of Exoproteome from some Filamentous Fungi Cultivated on Various Complex Polysaccharides

Cazy name	Main Enzyme name	*P. chrysosporium**				*Aspergillus niger*[§]		*Trichoderma reesei*[§]	*P.funiculosum*[#]
		HBA[1]	CLB[2]	NLB[3]	D	Glucose	Birchwood xylan	Lactose/xylose	(Rovabio EXL)
GH3	β-glycosidase/β-glucosidase	1	0	0	1	2	3	2	3
GH5	β-endomannase	4	0	0	4	2	2	1	2
GH6	exocellobiohydrolase	2	1	0	1	0	2	1	1
GH7	exocellobiohydrolase	4	2	0	3	0	2	2	1
GH10	Xylanase A	4	1	1	4	0	2	3	1
GH11	Endoxylanase (B,C,D)	1	0	4	1	1	0	0	4
GH12	Endoglucanase	2	1	0	2	2	1	1	1
GH28	Polygalacturonase	1	0	1	0	0	1	0	1
GH35	β-galactosidase	0	0	1	0	0	1	0	1
GH43	β-glycosidase	0	0	0	1	3	3	2	2
GH51	L-α-arabinofuranosidase	1	0	0	0	0	1	0	1
GH54	L-α-arabinofuranosidase	0	0	0	0	1	1	3	0
GH61	β-endoglucanase	1	1	1	2	1	1	0	8
GH62	L-α-arabinofuranosidase	0	0	0	0	0	2	0	2
GH74	Xyloglucanase	1	0	0	2	0	0	1	-
SWO	Expansin/swollenine	0	0	0	0	0	1	1	2
CE1	Acetylxylan esterase/feruloyl esterase	1	1	0	1	0	1	1	2
CE8	Pectinmethyl esterase	0	0	0	0	0	1	1	0
PL1	Pectate lyase	0	0	0	1	1	1	0	1
PL4	Rhamnogalacturonan lyase	0	0	0	1	0	1	0	1

*(42); [§] (39); [§](16); [#](15)

[1]HBA: Highley's basal medium supplemented with avicel; [2]CLB: carbon limited medium; [3]NLB: nitrogen limited medium

Components of the hemicellulolytic systems were also identified; including β-xylosidase, xylanases (XYN) and arabinofuranosidases (ABF), but the number and the abundance of these proteins were different between the two industrial strains. Only one xylanase (XYNIV) was expressed in RUT-C30, whereas CL847 strain also expressed two additional xylanases (XYNI and XYNII) at level higher than XYNIV. Also, while the two strains expressed three different ABF, ABFIII was 10-fold more abundant in CL847, and ABFI was slightly more expressed in RUT-C30 than in CL847. Thus, as for xylanases, this result may indicate that expression of these proteins is not co-regulated. Non-hydrolytic proteins CIPI and CIPII and one swollenin were also identified only in the culture filtrate from CL847. These proteins are likely to be important components of the cellulase cocktail produced by the industrial *T. reesei* CL847 for efficient lignocellulose degradation since they have a cellulose domain, and hence are thought to participate in cellulose degradation through the disruption of hydrogen bonds between polysaccharide chains. In RUT-C30, only CIPI is present, while other non-hydrolytic was comparable to that in CL847. Altogether, this comparative proteomic analysis underscored many unexpected differences between the exoproteome of two hypersecretory *T. reesei* strain that likely results from independent genetic selection from the same *T. reesei* QM6a followed by different adaptation to industrial conditions.

The exoproteome can also be influenced by the fermentation process even though the fungus is cultivated on the same carbon source, as has been reported for *Aspergillus oryzae* cultivated under submerged and solid-state culture conditions using wheat bran as the carbon source [27]. The comparative proteomic analysis between the two culture conditions not only showed that the amount of protein secreted by solid-state fermentation was about 4 to 6 times higher than that under the submerged conditions, but that the secretion of proteins was strongly influenced by the process. The 29 proteins identified from 85 spots on 2D gels were classified into four distinct groups. Group 1 consisted of extracellular proteins specifically produced in the solid-state growth conditions, including glucoamylase B (GH15), α-mannosidase and several proteins with unknown function. Group 2 contained extracellular proteins specifically produced under submerged condition, as glucoamylase A (GH15) and Xylanase G2 (from GH11 family), α and β-glucosidases (family GH3) and cellobiohydrolase (GH6,7). In Group 3 were secreted proteins present in both conditions, among which were several glycosyl hydrolases such as xylanase, xylosidase, L-α-arabinofuranosidase GH54 needed for degradation of hemicelluloses of wheat bran. Finally, group 4 consisted of proteins that were secreted in the medium under solid-state growth conditions but trapped at the cell wall in the submerged condition. Some insights on the molecular mechanism by which these conditions influence protein secretion were obtained from expression analysis of genes encoding some of the concerned proteins. It was found that *glaB* that encodes glucoamylase B was not expressed in submerged conditions, which explained its absence under this condition. On the other hand, *glaA* encoding glucoamylase A was expressed under both fermentation processes, but the protein was only found in the submerged conditions, indicating posttranslational modification of this protein under solid state condition. Also, α-amylase and β-glucosidase are likely to be subjected to posttranslational modification under solid state to account for export into the medium. Alternatively, this modification can be inhibited under submerged condition to trap these proteins at the cell wall, and prevent their secretion in the medium. The authors favored a role for cell wall in the postranslational control of these two proteins, based on the fact that a solid cell wall is required for submerged culture to cope with low external pressure, whereas the cell wall does not need to be so rigid under solid-state conditions, allowing an easier secretion of proteins. As a final remark, direct mass spectrometric analyses only identified a small set of proteins not solely because of technological limitations, but also because several spots isolated on the 2D-gels were not specifically induced by a given growth or process conditions, and thus were not included in the proteomic analysis. Nevertheless, the presence of these proteins indicates that there are many more secreted proteins whose function needs to be addressed.

Further Insight in Exproteome by Combination of Computational and Transcriptomic Analyses

The availability of several fungal genome sequences has allowed detailed computational analysis to refine the exoproteome of the filamentous fungi. Vanden Wymelenberg *et al.* [43] were among the first to exploit the genome sequence of the white-rod *P. chrysosporium* to predict the secretome of this fungus. A first version of genome annotation (v1.0) was used to predict 268 secreted proteins from 11,722 gene models using SignalP v.2, TargetP v.1 and TMHMM v.2 algorithms (http://www.cbs.dtu.dk/services/). These predictions were assessed by mass spectrometric analysis of extracellular proteins produced by *P. chrysosporium* cultivated on a standard cellulose-containing medium. A total of 182 peptide sequences were found to match 50 proteins from their gene models, from which only 25 out of 268 predicted proteins were present in the exoproteome. In addition, 32 glycosyl hydrolases implicated in cellulose/hemicelluloses degradation (*i.e.* cellobiohydrolase, xylanases, endoglucanases, acetyl xylan esterase, L-α-arabinofuranosidase, etc. see Table **4**) were detected as compared to 74 putative GH proteins predicted from the genome sequence (see Table **3**). This difference indicates significant disparity in gene regulation of GH enzymes that belong to the same families. As an example, the GH61 family has 14 members but only one of them has been found in the culture filtrate. In a second report, the improved gene annotation of this fungal genome led to a total of 10,048 gene models, from which a substantial increase of the secretome size to 769 proteins was predicted using the same algorithms (Table **2**) [42]. This version 2.1 of "computational secretome" was then used to analyze the exoproteome from carbon and nitrogen-limited growth conditions (this condition partially mimics ligninolytic condition) and from cellulose (avicel) cultures. As expected, the two conditions led to strikingly different protein patterns. Several hydrolytic enzymes including cellobiohydrolases, endoglucanases, β-glucosidase were detected only following culture of the filamentous fungus on cellulose medium, and not in carbon and nitrogen - depleted media. In contrast, several carbohydrate actives enzymes broadly characterized as hemicellulases (xylanases, exopolygalacturonase, β-galactosidase, α-mannosidase, acetyl xylan esterase) were present together with a large array of oxidative enzymes for lignin degradation (lignin peroxidase, manganese peroxidase, etc.), as well as several peptidases. More interestingly, 43 % of the predicted secretome contained proteins with still uncharacterized function. Since several of them harbored conserved cellulose binding modules, this reservoir of 'hypothetical' proteins should merit further investigation with respect to their role in fungal physiology in general, and in biomass degradation in particular.

A similar combined computational and mass spectrometric approach has been carried out to characterize the *A. niger* secretome. From 11,200 gene models in the *A. niger* ATCC1015 genome, a secretome size ranging from 691 to 881 secreted proteins was estimated, depending on the algorithm used [39]. This predicted secretome was experimentally verified by analysis of extracellular proteins produced by the fungus cultivated under 6 different media. A total of 222 proteins were identified, with 39 expressed under all conditions, and a total of 74 proteins expressed only under one condition. As already reported, several proteins were specifically expressed according to the growth substrate. This was notably the case for several pectinolytic enzymes, which were detected only when pectin was the carbon source. On the other hand, expression of most of the enzymes involved in hydrolysis of cellulose was more permissive since several of them were expressed under the six conditions tested (glucose, glycerol, sorbitol, birchwood xylan, pectin and locust bean gum) (Table **4** only reports data from glucose versus birchwood xylan). In addition, 90 % of the identified proteins by mass spectrometry possessed a signal peptide, which contrasts with the suggestion given above that several proteins secreted by *P. chrysosporium* do not necessarily possess secretion signals. This apparent discrepancy actually reveals that the sequence-based analysis of secretomes ultimately depends on the reliability of the gene model predictions and on the quality of the sequence. Proteomic analysis can provide experimental evidence to verify these gene models and in some cases, correct them.

The functionality of the secretome can be further explored by transcriptomic methods using microarrays or massively parallel pyrosequencing. These technologies are complementary to *in silico* analysis, as they could indicate whether or not these gene models are expressed and in a quantitative manner with the latter method [4]. It is also complementary to proteomic analysis because gene expression analysis can overcome, in part, the limitation of MS-based protein identification for lowly expressed proteins. However, this relationship is only true if one assumes a good correlation between transcripts and protein levels, which has been rarely demonstrated, even in yeast [10]. Nonetheless, this genome-transcriptome approach has been undertaken to investigate the potential of *P. chrysosporium* for wood degradation [32]. The transcriptome was determined by massive parallel pyrosequencing of an expression library from the white-rot fungus grown on a solid substrate, namely the red oak. This study showed that only cDNA encoding endoxylanase (GH10, GH11), acetyl xylan esterase (CE1) and endomannanase (GH5), among a set of a larger number of genes encoding putative cellulose and hemicellulose degradation enzymes, were detected. Interestingly, these transcriptomic data supported a previous proteomic analysis of exoproteome from this species cultivated under this condition, since only the same subset of glycosyl hydrolase enzymes were identified in the culture filtrate [33]. In addition, this analysis showed that this fungus put a large investment into producing peroxidases at very high level, and revealed an impressive list of upregulated genes encoding potential secreted proteins whose function is still completely unknown.

A similar approach has been followed up by Vanden Wymelenberg *et al.* [41] who investigated the enzymatic potential of *P. chrysosporium* for cellulose and lignocellulose degradation by microarray analysis and confronted the expression profiles with proteome analysis of culture filtrate by LC-MS/MS. As expected, they found that the larger category of upregulated genes (*i.e.* 70 out of 535 upregulated genes) in cellulose and lignocellulose media were those encoding Carbohydrates Active enzymes. From this analysis, they also pointed out a wide heterogeneity in expression of genes that belong to a same family. For instance, only 6 out of the 13 members of the GH61 family were expressed under cellulose conditions. This condition was also specific in causing upregulation of more than 190 genes encoding proteins of unknown function. Interestingly, the sequence of one-third of these upregulated genes featured easily identifiable secretion sequences, and peptides corresponding to 54 gene models were identified in the culture filtrate from cellulose-growing fungus. However, some of these secreted proteins did not show canonical secretion signals, although they exhibited other features such as GPI-anchor, or a carbohydrate binding module. It is likely that these proteins specifically expressed in a cellulose/lignocellulose medium may directly participate in the degradation of these complex polysaccharides.

Biodegradation of cellulose and hemicelluloses biodegradation does not necessarily involve cellulolytic systems. It can also involve the attack of cellulose by hydroxyl free radicals in conjunction with a limited set of endoglucanase and β-glucosidases. This mechanism has long been proposed to contribute to polysaccharide depolymerisation by some brown rot filamentous fungi (reviewed in [2]), and is now supported by genome-transcriptome and secretome analysis of the brown-rot *P. placenta* [22]. While the genome sequencing of this fungus predicts a similar set of genes encoding CAZymes as in other filamentous fungi (see Table **2**), it reveals the absence of conventional cellulase encoding genes. In addition, transcriptome analysis showed upregulation of a limited set of endoglucanases (GH3, GH12 and GH61), and few hemicellulase encoding genes (endoxylanases, β-xylosidase, laminarinase, and one L-α-arabinofuranosidase). On the other hand, the expression profile of *P. placenta* on cellulose showed the upregulation of several genes involved in iron transport, quinone reductase, cellulose dehydrogenase and low molecular glycoproteins that can act as iron reductases, suggesting that cellulolytic conditions might activate high affinity iron-uptake system, which in turn increases intracellular levels of Fe(II) that reacts with H_2O_2 to produce radical hydroxyl by the Fenton reaction.

CONCLUSIONS AND PERSPECTIVES

A combination of genomics, transcriptomics and proteomics is the most straightforward strategy to capture the great enzymatic potential of filamentous fungi that is needed for efficient biomass degradation. While initial proteomic data obtained by advanced mass spectrometry techniques showed that these microorganisms can produce a broad range of enzymes with different and complementary activities, specifically tailored for use and consumption of their growth substrates, genome sequencing has underscored the presence of a larger repertoire of genes encoding putative Carbohydrate Active enzymes. Moreover, whole genome expression analysis from filamentous fungi cultivated on cellulose medium revealed the upregulation of a large set of genes encoding an impressive array of potential secreted proteins with yet unknown functions. Altogether, this systems biology approach suggests that these fungi are endowed with a much more complex enzymatic potential than expected for optimized utilization of complex substrates. These genomic-transcriptomic data raises several new avenues for a better understanding of the mechanism by which these filamentous fungi can degrade biomass. An urgent question is to validate biochemically the large repertoire of these putative Carbohydrate Active proteins that have been identified in filamentous fungal genomes. It is clear that gene redundancy found in most of the CAZy families is intriguing and raises the question of whether the expression of these genes is tightly dependent on growth conditions; or that the encoded proteins that belong to the same family have the same catalytic function. Identification of the function of the large set of hypothetical secreted proteins in biomass degradation is another burning issue. In most instances, heterologous enzyme production for biochemical studies and genetic suppression will be necessary to achieve these goals, even though these approaches are technically challenging particularly on a high-throughput basis. Regulation of gene expression is another research area that is still in infancy in filamentous fungi, probably because these microorganisms are less amenable to genetic analysis than yeasts. However, the development of post-genomic technologies combined with powerful bioinformatic tools now offers a means to investigate these questions. Along this line, Couthino *et al.* [6] have pointed out significant differences between three *Aspergilli* species (*A. niger*, *A. nidulans* and *A. oryzae*) for the presence of putative regulatory sequences in the promoters of orthologues and showed a correlation only between the presence of the XlnR binding site and xylose induction in *A. niger*. This result indicates that either the XlnR consensus sequence is not exactly the same between species, or more likely that there are many other factors regulating gene expression, whose action is linked to individual preference for substrates and role in the natural biotope. Understanding the remarkable secretion capacity of these filamentous fungi is also an important issue towards improving homologous and heterologous protein production. Quite surprisingly, very little mechanistic insight has been obtained from the genome sequence analysis. This could mean that this remarkable capacity for protein secretion lies rather at the posttranslational level and/or implies different membrane trafficking system than in less-efficient secretory organism such as the yeast *S. cerevisiae*.

The perspective is thus to follow up into this Systems Biology approach combining genomics, bioinformatics, transcriptomics and proteomics, and to complement this global approach by detailed biological and biochemical validation of the identified proteins. Results gathered from this approach should help deciding the optimal strategy for biomass degradation technology. Thus, data will enable a more realistic assessment from a scientific and economical viewpoint whether to create a superior filamentous fungal species that can express at high level the desired enzymes by genetic engineering or to reconstitute an optimized enzyme cocktail from large scale heterologous proteins production by fermentation of recombinant microorganisms on low cost carbon sources.

REFERENCES

[1]   Abbas A, Koc H, Liu F, Tien M. Fungal degradation of wood: initial proteomic analysis of extracellular proteins of *Phanerochaete chrysosporium* grown on oak substrate. Curr Genet 2005; 47: 49-56.

[2]   Baldrian P, Valaskova V. Degradation of cellulose by basidiomycetous fungi. FEMS Microbiol Rev 2008; 32(3): 501-21.

[3]   Bendtsen JD, Nielsen H, von Heijne G, Brunak S. Improved prediction of signal peptides: SignalP 3.0. J Mol Biol 2004; 340:783-95.

[4]   Brenner S, Johnson M, Bridgham J, *et al.* Gene expression analysis by massively parallel signature sequencing (MPSS) on microbead arrays. Nat Biotechnol 2000; 18: 630-4.

[5]   Cantarel BL, Coutinho PM, Rancurel C, Bernard T, Lombard V, Henrissat B. The Carbohydrate-Active EnZymes database (CAZy): an expert resource for Glycogenomics. Nucleic Acids Res 2009; 37:D233-8.

[6]   Coutinho PM, Andersen MR, Kolenova K, *et al.* Post-genomic insights into the plant polysaccharide degradation potential of *Aspergillus nidulans* and comparison to *Aspergillus niger* and *Aspergillus oryzae*. Fungal Genet Biol 2009; 46(1):S161-9.

[7]   Cuomo CA, Guldener U, Xu JR, *et al.* The *Fusarium graminearum* genome reveals a link between localized polymorphism and pathogen specialization. Science 2007; 317: 1400-2.

[8]   Dean RA, Talbot NJ, Ebbole DJ, *et al.* The genome sequence of the rice blast fungus *Magnaporthe grisea*. Nature 2005; 434: 980-6.

[9]   Emanuelsson O, Brunak S, von Heijne G, Nielsen H. Locating proteins in the cell using TargetP, SignalP and related tools. Nat Protoc 2007; 2: 953-71.

[10]   Futcher B, Latter GI, Monardo P, McLaughlin CS, Garrels JI. 1999. A sampling of the yeast proteome. Mol Cell Biol 1999; 19: 7357-68.

[11]   Galagan JE, Calvo SE, Borkovich KA, *et al.* The genome sequence of the filamentous fungus *Neurospora crassa*. Nature 2003; 422: 859-68.

[12]   Galagan JE, Calvo SE, Cuomo C, *et al.* Sequencing of *Aspergillus nidulans* and comparative analysis with *A. fumigatus* and *A. oryzae*. Nature 2005; 438: 1105-15.

[13]   Goffeau A, Barrell BG, Bussey H, *et al.* Life with 6000 genes. Science 1996; 274(5287): 546, 563-7.

[14]   Greenbaum D, Luscombe NM, Jansen R, Qian J, Gerstein M. Interrelating different types of genomic data, from proteome to secretome: 'oming in on function. Genome Res 2001; 11: 1463-8.

[15]   Guais O, Borderies G, Pichereaux C, *et al.* Proteomics analysis of "Rovabiot Excel", a secreted protein cocktail from the filamentous fungus *Penicillium funiculosum* grown under industrial process fermentation. J Ind Microbiol Biotechnol 2008; 35: 1659-68.

[16]   Herpoel-Gimbert I, Margeot A, Dolla A, *et al.* Comparative secretome analyses of two *Trichoderma reesei* RUT-C30 and CL847 hypersecretory strains. Biotechnol Biofuels 2008; 1: 18.

[17]   Kall L, Krogh A, Sonnhammer EL. Advantages of combined transmembrane topology and signal peptide prediction--the Phobius web server. Nucleic Acids Res 2007; 35: W429-32.

[18]   Le Crom S, Schackwitz W, Pennacchio L, *et al.* Tracking the roots of cellulase hyperproduction by the fungus *Trichoderma reesei* using massively parallel DNA sequencing. Proc Natl Acad Sci USA 2009; 106: 16151-6.

[19]   Leonowicz A, Matuszewska A, Luterek J, *et al.* Biodegradation of lignin by white rot fungi. Fungal Genet Biol 1999; 27: 175-85.

[20]   Machida M, Asai K, Sano M, *et al.* Genome sequencing and analysis of *Aspergillus oryzae*. Nature 2005; 438: 1157-61.

[21]   Martinez D, Berka RM, Henrissat B, *et al.* Genome sequencing and analysis of the biomass-degrading fungus *Trichoderma reesei* (syn. Hypocrea jecorina). Nat Biotechnol 2008; 26: 553-60.

[22]   Martinez D, Challacombe J, Morgenstern I, *et al.* Genome, transcriptome, and secretome analysis of wood decay fungus *Postia placenta* supports unique mechanisms of lignocellulose conversion. Proc Natl Acad Sci USA 2009; 106: 1954-9.

[23]   Martinez D, Larrondo LF, Putnam N, *et al.* Genome sequence of the lignocellulose degrading fungus *Phanerochaete chrysosporium* strain RP78. Nat Biotechnol 2004; 22: 695-700.

[24]   Miettinen-Oinonen A, Paloheimo M, Lantto R, Suominen P. Enhanced production of cellobiohydrolases in *Trichoderma reesei* and evaluation of the new preparations in biofinishing of cotton. J Biotechnol 2005; 116: 305-17.

[25]   Nierman WC, Pain A, Anderson MJ, *et al.* Genomic sequence of the pathogenic and allergenic filamentous fungus *Aspergillus fumigatus*. Nature 2005; 438: 1151-6.

[26]   Nombela C, Gil C, Chaffin WL. Non-conventional protein secretion in yeast. Trends Microbiol 2006; 14: 15-21.

[27]   Oda K, Kakizono D, Yamada O, Iefuji H, Akita O, Iwashita K. Proteomic analysis of extracellular proteins from *Aspergillus oryzae* grown under submerged and solid-state culture conditions. Appl Environ Microbiol 2006; 72:3448-57.

[28]   Paper JM, Scott-Craig JS, Adhikari ND, Cuomo CA, Walton JD. Comparative proteomics of extracellular proteins *in vitro* and in planta from the pathogenic fungus *Fusarium graminearum*. Proteomics 2007; **7**: 3171-83.

[29]   Pel H J, de Winde JH, Archer DB, *et al.* Genome sequencing and analysis of the versatile cell factory *Aspergillus niger* CBS 513.88. Nat Biotechnol 2007; 25: 221-31.

[30]   Ratto M, Ritschkoff A, Viikari L. The effect of oxidative pretreatment on cellulose degradation by *Poria placenta* and *Trichoderma reesei*. Appl Microbiol Biotechnol 1997; 80: 53-7.

[31]   Ravalason H, Jan G, Molle D, *et al.* Secretome analysis of *Phanerochaete chrysosporium* strain CIRM-BRFM41 grown on softwood. Appl Microbiol Biotechnol 2008; 80: 719-33.

[32]   Sato S, Feltus FA, Iyer P, Tien M. The first genome-level transcriptome of the wood-degrading fungus *Phanerochaete chrysosporium* grown on red oak. Curr Genet 2009; 55: 273-86.

[33]   Sato S, Liu F, Koc H, Tien M. Expression analysis of extracellular proteins from *Phanerochaete chrysosporium* grown on different liquid and solid substrates. Microbiology 2007; 153: 3023-33.

[34]   Shoseyov O, Shani Z, Levy I. Carbohydrate binding modules: biochemical properties and novel applications. Microbiol Mol Biol Rev 2006; 70: 283-95.

[35]   Soanes DM, Alam I, Cornell M, *et al.* Comparative genome analysis of filamentous fungi reveals gene family expansions associated with fungal pathogenesis. PLoS One 2008; 3:e2300.

[36]   Soanes DM, Richards TA, Talbot NJ. Insights from sequencing fungal and oomycete genomes: what can we learn about plant disease and the evolution of pathogenicity? Plant Cell 2007; 19: 3318-26.

[37]   Suarez MB, Sanz L, Chamorro MI, *et al.* Proteomic analysis of secreted proteins from *Trichoderma harzianum*. Identification of a fungal cell wall-induced aspartic protease. Fungal Genet Biol 2005; 42: 924-34.

[38]   Tjalsma H, Bolhuis A, Jongbloed JD, Bron S, van Dijl JM. Signal peptide-dependent protein transport in *Bacillus subtilis*: a genome-based survey of the secretome. Microbiol Mol Biol Rev 2000; 64: 515-47.

[39]   Tsang A, Butler G, Powlowski J, Panisko EA, Baker SE. Analytical and computational approaches to define the *Aspergillus niger* secretome. Fungal Genet Biol 2009; 46(1): S153-60.

[40]   van den Berg MA, Albang R, Albermann K, *et al.* Genome sequencing and analysis of the filamentous fungus *Penicillium chrysogenum*. Nat Biotechnol 2008; 26: 1161-8.

[41]   Vanden Wymelenberg A, Gaskell J, Mozuch M, *et al.* Transcriptome and secretome analyses of *Phanerochaete chrysosporium* reveal complex patterns of gene expression. Appl Environ Microbiol 2009; 75: 4058-68.

[42]   Vanden Wymelenberg A, Minges P, Sabat G, *et al.* Computational analysis of the *Phanerochaete chrysosporium* v2.0 genome database and mass spectrometry identification of peptides in ligninolytic cultures reveal complex mixtures of secreted proteins. Fungal Genet Biol 2006; 43: 343-56.

[43]   Vanden Wymelenberg A, Sabat G, Martinez D, *et al.* The *Phanerochaete chrysosporium* secretome: database predictions and initial mass spectrometry peptide identifications in cellulose-grown medium. J Biotechnol 2005;118: 17-34.

[44]   Yang J, Li C, Wang Y, *et al.* Computational Analysis of signal peptide-dependent secreted proteins in *Saccharomyces cerevisiae*. Agric Sci China 2006; 5: 221-7.

[45]   Zorn H, Peters T, Nimtz M, Berger RG. The secretome of *Pleurotus sapidus*. Proteomics 2005; 5: 4832-8.

CHAPTER 4

Structural Biology of Fungal Multicopper Oxidases

Francisco J. Enguita*

Unidade de Biologia Celular. Instituto de Medicina Molecular. Faculdade de Medicina. Universidade de Lisboa. Av. Prof. Egas Moniz. 1649-028 Lisboa. Portugal. E-mail: fenguita@fm.ul.pt

Abstract: Copper containing proteins are widespread in nature, ranging from humans to simple prokaryotes. They are involved in several functions related with copper homeostasis, transport and energy metabolism. Multicopper oxidases are a well characterized group of copper-containing proteins, characterized by their ability to employ the redox properties of the copper atoms to catalyze substrate oxidations. The multicopper oxidases have been classified in three distinct families according to their fold and properties: (I) nitrite reductase-type, (II) laccase-type and (III) ceruloplasmin-type. In fungi, multicopper oxidases showed versatile metabolic properties, being involved in reactions such as polymerization of phenolic acids to constitute lignin, and the oxidative degradation of xenobiotic compounds. In this review we will analyze in detail the structural properties of two main groups of fungal multicopper oxidases, laccases and tyrosinases, in order to establish solid structure-activity relationships among all their members. Structural features of multicopper oxidases from fungal origin will be also related with the substrate specificity of the enzymes and their possible applications in biotechnological processes.

INTRODUCTION

Transition metals are common constituents of enzymes, frequently incorporated as cofactors that help in the catalysis of chemical reactions. Examples of enzymes containing metals as cofactors include electron transport proteins, redox-enzymes, and some proteins involved in the catalysis of complex reactions such as photosynthesis, nitrogen fixation or nucleotide biosynthesis. In spite of the presence of some "exotic" elements such as cobalt or molybdenum in some enzymes [52, 61, 68, 71], the most common transition elements in proteins are iron, zinc and copper [85]. Because of their inherent reduction potentials and coordination chemistry, the only metals present in biological systems and able to catalyze redox reactions are iron and copper [36]. One of the most important properties of iron and copper is that they are also able to bind and manage gas substrates such as oxygen [6, 23, 36].

Copper and iron availabilities in biological systems have evolved during Earth history, and this fact has been studied in comparison with other planetary geological systems [18]. Geological conditions of primitive Earth firstly favored the availability of Fe(II) and sulphur compounds before the invention of the photosynthesis.

Reduced iron-sulphur (sulphides) compounds were more abundant and available to biological systems in the primitive Earth because these compounds are intrinsically more soluble than the corresponding copper derivatives. The photosynthetic life forms slowly begin to produce oxygen, and the availability of this gas in the atmosphere allowed the oxidation of copper releasing it in the form of soluble copper oxides. The availability of soluble copper allowed microorganisms to incorporate it and to perform new chemical reactions based on this element, that are currently represented by enzymes involved in redox reactions [37, 43].

The copper properties directly related with its catalytic efficiency and prevalence in biological systems are the chemical stability of its redox-interchangeable main ionic forms, its ability to bind oxygen and its affinity for functional groups that occur in proteins. Accordingly, copper-containing enzymes are found in electron transfer pathways, such as respiration and photosynthesis, and also in proteins that bind oxygen, such as oxidases and oxygen transporters. Because of its role in fundamental biochemical pathways, copper is also an essential micronutrient for all the cells. However, the same properties that make copper a useful cofactor in biochemical redox reactions or oxygen chemistry can wreak havoc within the cell when intracellular copper content is excessive, and accordingly, organisms have developed methods for avoiding its toxicity [37, 43, 70, 82].

*Address correspondence to Franciso J. Enguita: Unidade de Biologia Celular. Instituto de Medicina Molecular. Faculdade de Medicina. Universidade de Lisboa. Av. Prof. Egas Moniz. 1649-028 Lisboa. Portugal. E-mail: fenguita@fm.ul.pt

Within proteins, copper ions are directly coordinated by aminoacids (usually sulphur containing aminoacids and histidine), and are not usually present in complex prosthetic groups. Several copper centers have been described in the literature and will be reviewed in this article in the context of each analyzed group of enzymes [36, 76].

Fungi are very versatile microorganisms, able to be adapted to almost all the natural environments. Their metabolic abilities have been target of increased attention since the last half a century, including the production of bioactive secondary metabolites [7, 33, 98, 115], toxins [34, 90, 95, 114], and enzymes [2, 3, 14, 22, 25, 69]. In the field of fungal enzymes, multicopper oxidases have become a subject of great interest because of their intrinsic properties that allow their application in a wide range of fields. Oxidases containing multiple copper centers are versatile biocatalysts able to catalyze the formation of free radicals from phenolic substrates without the production of reactive oxygen species [75, 104]. In nature, these enzymes are related with de-polymerization of complex phenolic derivatives such as lignin [42, 58, 97], degradation of xenobiotics [5, 97], and biosynthesis of cell structures [46, 75].

In this review we will analyze the structural properties of several groups of fungal multicopper oxidases, focusing our attention in their properties and distribution among different species of fungi.

COPPER-CONTAINING PROTEINS

Copper proteins contain one or more copper atoms coordinated directly by aminoacids that stabilize the coordination sphere of the metal ion. Interestingly, copper coordination within proteins is extremely diverse and the metal centers in the copper proteins can be classified into several types:

Type I copper centers are characterized by a single copper atom coordinated by two histidine residues, a cysteine residue in a trigonal planar structure, and a variable axial ligand. Among the proteins containing a single Type-I copper center there are three main families. In the family I (also named as "Class" I) (e.g. amicyanin, plastocyanin and pseudoazurin) the variable axial ligand of the copper atom is a methionine, whereas aminoacids other than methionine in the same coordination position (e.g. glutamine) give rise to the family II of mononuclear copper proteins [9, 36, 77]. Azurins contain the third type of type-I copper centers (family III): they contain a different second axial ligand (a carbonyl group of a glycine residue) [9]. Proteins containing only type-I copper centers are usually called "cupredoxins", and show similar three-dimensional structures, relatively high reduction potentials (> 250 mV), and strong absorption near 600 nm (due to the electronic transfer S→Cu), which usually gives rise to a blue color. Cupredoxins are therefore often called "blue copper proteins", however this fact could be sometimes misleading, since some type-I copper centers also absorb around 460 nm and are therefore green. When studied by EPR spectroscopy, type-I copper centers show only small hyperfine splitting in the parallel region of the spectrum if compared with normal coordination copper atoms [9, 36, 88].

Type II copper centers exhibit a square planar coordination mainly ensured by N or N/O atom ligands belonging to histidine and glutamate residues [23, 64]. The axial EPR spectrum showed a copper hyperfine splitting in the parallel region similar to that observed in regular copper coordination compounds [64]. Since no sulfur ligation is present, the optical spectra of these centers lack distinctive features, mainly the characteristic absorption at 600 nm which is typical from type-I copper centers. Type-II copper centers appear in complex enzymes, where they assist in oxidations or oxygenations, and they are frequently associated to type-III centers (eg. : laccases and other phenol oxidases) [1, 23, 38, 64].

Type III copper centers are binuclear centers consisting of two copper atoms, each coordinated by three histidine residues [38]. These proteins exhibit no EPR signal due to strong antiferromagnetic coupling (*i.e.* spin pairing) between the two $S = 1/2$ metal ions due to their covalent overlap with a bridging ligand. These centers are present in some oxidases and oxygen-transporting proteins acting normally in coordination with type-II copper centers (e.g. tyrosinases and laccases) [38].

There are other less common copper coordination schemes in proteins such as the binuclear copper A centers found in cytochrome *c* oxidase and nitrous-oxide reductase, copper B centers that are also present in cytochrome *c* oxidase, and the tetranuclear copper Z center (Cu_Z) is found in nitrous-oxide reductase [53, 80, 85, 109].

Protein domains involved in copper coordination showed a common fold composed of eight beta strands forming a beta-barrel and called "cupredoxin" [74]. It has been assumed that cupredoxin domains have evolved to sequester copper atoms from the environment, since this metal is toxic in its free form [43]. Proteins containing cupredoxin domains are very abundant in nature and could be classified in Mono-domain cupredoxins, such as amicyanin, plastocyanin, pseudoazurin, plantacyanin, azurin, auracyanin, rusticyanin, stellacyanin, and mavicyanin; Multi-domain cupredoxins, such as nitrite reductase (2 domains of this fold), multicopper oxidase CueO, spore coat protein A (CotA) from gram positive bacteria, ascorbate oxidases (3 domains of this fold), laccases (3 domains of this fold), ceruloplamin (6 domains of this fold), and coagulation factor V; and other proteins with different domain combinations such as Red copper protein nitrocyanin and the C-terminal of nitrous oxide reductase, Quinol oxidase and the periplasmic domain of cytochrome *c* oxidase subunit II, and the N-terminal domain of protein arginine deaminase Pad4, which is related to cupredoxin but lacks the metal-biding site. Cupredoxin domains are considered as the building blocks that nature has used to construct several types of proteins [74].

LACCASES

Introduction

Laccases (benzenediol: oxygen oxidoreductase, EC 1.10.3.2) are multi-copper oxidases widely distributed in nature ranging from prokaryotes [16, 17] to complex eukaryotes [5, 69]. Laccase activity was firstly discovered in the exudates of the Japanese lacquer tree *Rhus vinifera* at the end of the 19th century [113], however these enzymes have been of increasing interest for research applications, mainly because their implications in the degradation of highly polymeric substances such as lignin [100] and xenobiotics [50, 59]. The majority of the laccases described in the literature have been isolated and characterized from higher fungal species. Laccases have been found in white-rot fungi, in saprophytic and phytopathogenic fungi, and also in agarics [60, 70, 86].

Laccases are extremely efficient oxidases, able to catalyze monoelectronic oxidations from a reducing substrate (normally phenolic compounds) generating free radical species by using the favorable redox properties of the copper atoms. Electrons extracted from the substrate are stored in a complex core structure constituted by four copper atoms, and finally transferred to oxygen to produce two water molecules (Fig. **1**). This reaction mechanism allows oxidizing a great variety of different substrates without producing a hydrogen peroxide intermediate. Each laccase molecule is able to extract and store four electrons before transferring them to the final acceptor, molecular oxygen, in a high yield mechanism. Depending on the particular characteristics of the laccase (redox potential of the copper atoms, isolectric point, structure of the substrate binding site, etc ...) they show wider or narrower substrate specificities. However we can define laccases as multicopper oxidases able to oxidize polyphenols, methoxylated phenols, aromatic diamines and other aromatic compounds, but unable to oxidize tyrosine as tyrosinases do.

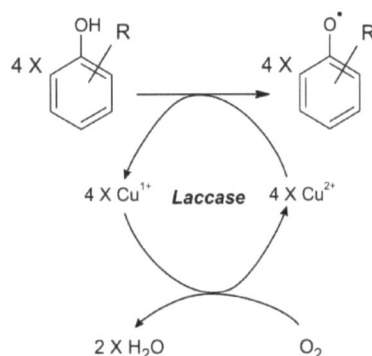

Figure 1: General reaction mechanism of laccases. Four monolectronic oxidations are catalyzed from a substituted (R) phenolic substrate generating the corresponding free radicals. Electrons are then transferred and stored by using the four copper atoms within the protein, and finally transferred to molecular oxygen to produce two water molecules.

In plants, laccases are mainly involved in anabolic oxidations related with the biosynthesis of polyphenolic polymers such as lignin whereas in fungi have more wide roles, being related with morphogenesis [29, 100], stress response [57], and degradation of xenobiotic compounds including those from vegetal origin [56, 100]. More in detail, the natural role of fungal laccases in the degradation of polymers from vegetal origin has become a controversial issue. Monoelectronic oxidations catalyzed by fungal laccases using small phenolic compounds as substrates to generate free radicals are not enough to guarantee depolymerization of high molecular weight structures such as lignin, because the intrinsic ability of phenolic free radicals to combine again producing polymeric species [70].

In the last two decades more than 100 different fungal laccases have been purified and isolated from their natural sources. They are mainly monomeric proteins with a molecular weight from 50 to 100 kDa, an acidid isoelectric point ranging from 2.5 to 4.5 units, and an optimum temperature around 50 °C, as described and compiled in BRENDA database [13].

In fungi, and due to the properties of their substrates, laccases are expected to be mainly extracellular enzymes. However the majority of the fungal laccases, especially those found in white-rot fungi are produced in both of the forms being possible to detect intra and extracellular laccase activity [104, 110]. Like most of the fungal extracellular proteins, laccases are glycoproteins. Their glycosylation extend ranges from 10 to 25% [86]. It has been proposed that in addition to the structural role, glycosylation could be also involved in the protection of laccase from proteolytic degradation [110]. The compulsory requierement for enzymatic activity of post-translational modifications in some fungal laccases has prevented their wide use in biotechnological applications [99].

Structural Analysis of Fungal Laccases

Tertiary structure of fungal laccases usually comprises three cupredoxin-like domains, an eight- stranded Greek key beta-barrel, connected by flexible loops [74]. Until now, only ten structures of fungal tri-domain multicopper oxidases have been determined by X-ray crystallography at different resolution ranges and the corresponding atomic coordinates deposited in the PDB database (Fig. **2**). They include mainly laccases from species belonging to the phylum *basidiomycota*, and also two multicopper oxidases from ascomycetes: *Melanocarpus albomyces* (PDB code: 1GW0) and the protein Fet3p from the yeast *Saccharomyces cerevisiae* (PDB code: 1ZPU). A Fet3p protein is a laccase-like protein involved in iron metabolism [19, 55].

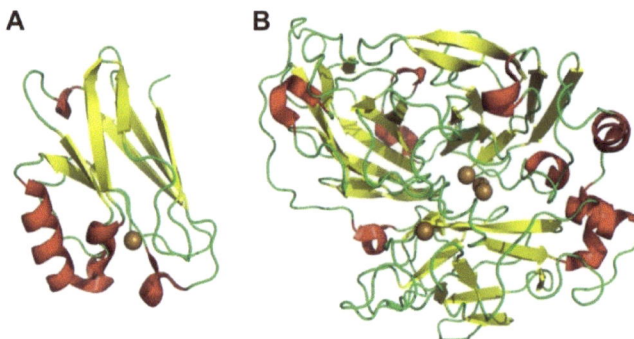

Figure 2: Structural representation of monodomain cupredoxin (A) and tri-domain laccase (B) from *C. cinereus* (PDB code: 1KYA). Combination of cupredoxin domains in laccase allowed the protein to bind more copper atoms in the cavity formed by the interfaces between domains. Structures have been colored by secondary structural elements, and the copper atoms represented as brown balls. Figure has been prepared with Pymol [21].

Sequences of proteins containing cupredoxin-like domains normally showed low homology in despite of their conserved structures. Structure based-sequence comparison of fungal laccases with known tridimensional structures is represented in Fig. **3**. The alignment clearly showed the presence of three well defined structural domains with a different degree of sequence homology. The first cupredoxin domain comprises approximately the first 150 aminoacids of the protein, and contains two conserved motifs, TSXHXHG and GTXWYHSH, involved in binding of copper atoms of the trinuclear T2/T3 copper center [64, 102, 107]. This domain is connected to the second cupredoxin fold by a short loop, which shows low sequence homology amount the analyzed proteins, specially in the multicopper oxidases from ascomycetes (PDB codes : 1GW0 and 1ZPU).

The second cupredoxin domain comprises approximately aminoacids from 150 to 330, showing a lower degree of sequence homology among the fungal proteins. This domain contains to histidine residues involved in copper binding in T2/T3 copper center. The last cupredoxin-like domain is well conserved in all the analyzed enzymes, with the exception of the multicopper oxidases from ascomycetes which showed a long sequence stretch in the middle of the domain that divides it in two symmetric halves (Fig. **3**). This long sequence insertion could potentially act as a protein hinge that will increase the flexibility of the domain packing of the enzymes. In the last cupredoxin domain, there are also two conserved sequence motifs, HPXHLHGH and NPGXWXXHCHI, involved in copper coordination of the remaining atoms of the T2/T3 center and the T1 blue copper center [9, 37, 64, 88].

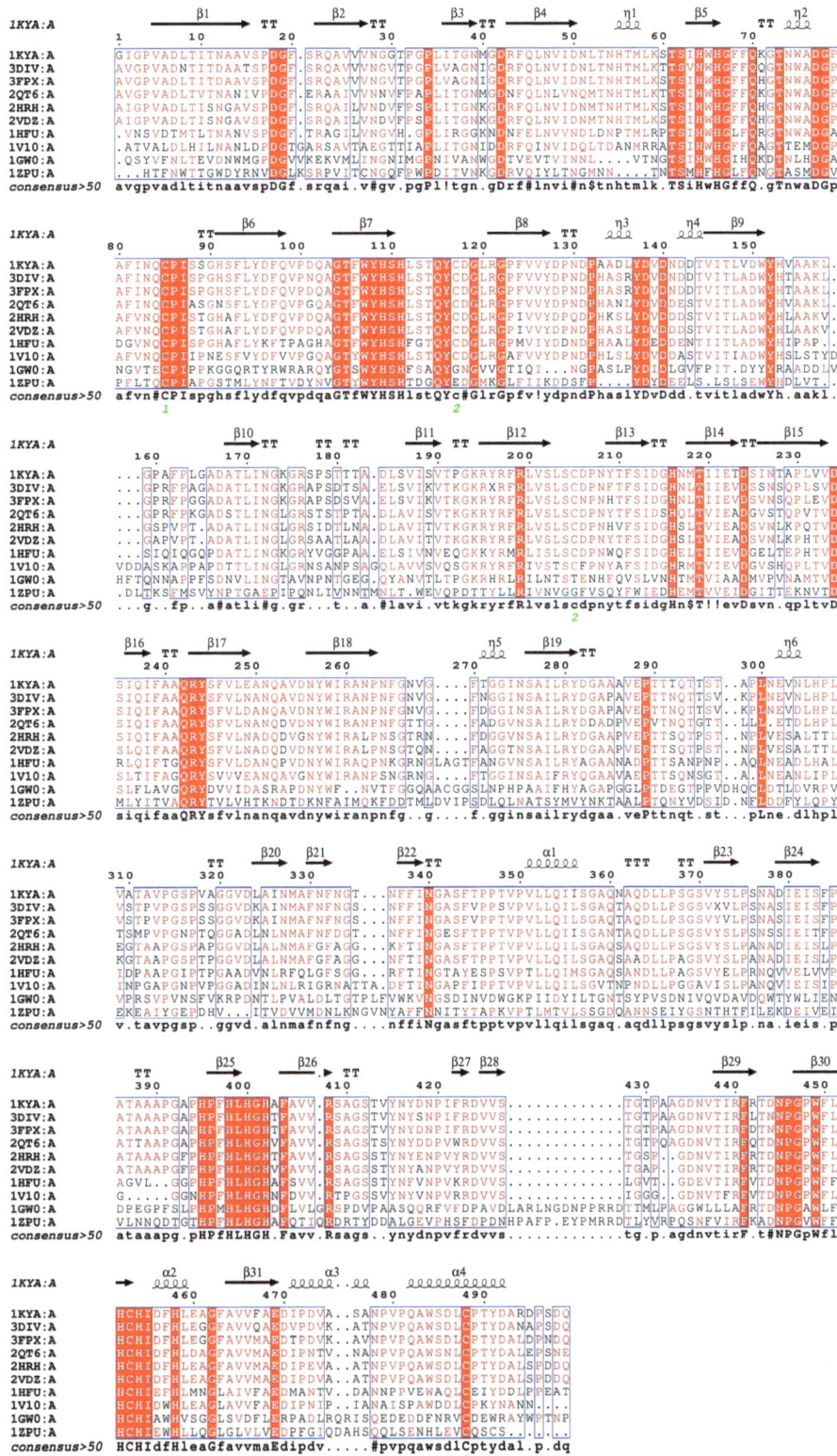

Figure 3: Structure-based sequence alignment of tri-domain fungal multicopper oxidases performed by Multiprot [94]. Secondary structure elements *Trametes versicolor* laccase (PDB code 1KYA) are represented on the top of the alignment. Graph was prepared with ESPript [35]. 1KYA: laccase from *T. versicolor*; 1GW0: laccase from *Melanocarpus albomyces*; 1ZPU: Fet3p protein from *Scccharomyces cerevisiae*; 1V10: laccase from *Rigidoporus lignosus*; 1HFU: laccase from *Coprinus cinereus*; 3DIV: laccase from *Cerrena maxima*; 3FPX: laccase from *Trametes hirsuta*; 2QT6: laccase from *Lentinus tigrinus*; 2HRH: laccase from *Trametes trogii* and 2VDZ: laccase from *Coriolopsis gallica*.

Phylogenetic analysis based on the structure alignment (Fig. **4**) showed an evident clustering of laccases from basidiomycetes aside from laccase from *M. albomyces* and the Fet3p protein from *S. cerevisiae*. This is consistent not only with the taxonomic classification of different species but also with the intrinsic characteristics of each enzyme. Laccase from *M. albomyces* showed a higher molecular weight (80 kDa) in comparion with the other analyzed members of the family, and also a relative higher optimum temperature (70-75 ºC) [38, 45]. Fet3p is a laccase-like protein involved in iron metabolism, however it shares folding and structure with laccases, being a possible specialized evolution of these enzymes [55, 91, 103].

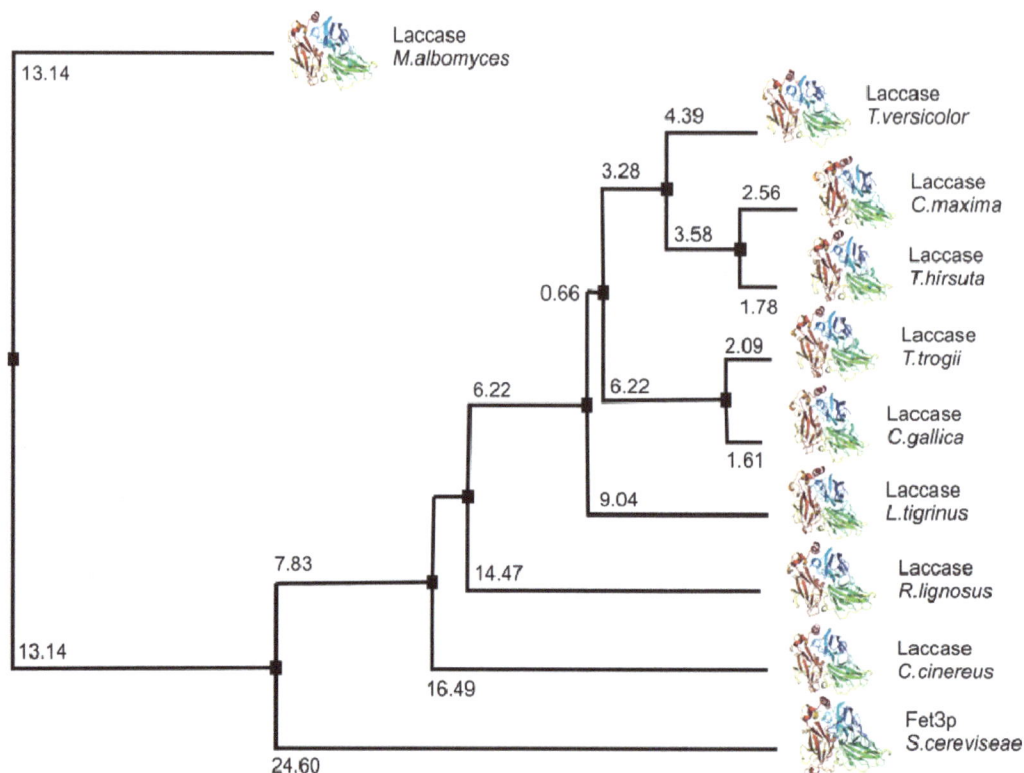

Figure 4: Phylogenetic tree constructed with data extracted from the structural comparison of fungal laccases and laccase-like proteins (Fet3p) by the FATCAT algorythm [106,112]. Distances between nodes are represented in the tree by numbers. Tree was prepared with Bosque software [83].

Mononuclear Copper Center T1

Type I copper center is responsible for an intense visible absortion at 600 nm that confers blue color to pure laccase samples, and is mainly located close to the protein surface (Fig. **5**). This center, that catalyzes the extraction of a single electron from a substrate, is mononuclear and contains a copper atom coordinated by three main residues, two histidines and a cysteine. In some laccases a four coordinating ligand is also present, frequently a methionine, however the function of this group in the electron transfer mechanism is still under discussion because the electron-donating ability of the thioether group is considerable lower when compared with the cysteine sulphur group or the nitrogens within the histidine imidazole rings [47, 48, 88, 105]. In fungal laccases is also common to find an aromatic aminoacid such as phenylalanine or tyrosine, or an aliphatic chain such as leucine or isoleucine as a fourth ligand [31, 62, 69, 81].

Among the fungal laccases of already know tridimensional structure, type-1 copper center is coordinated by two histidines and a cysteine, and a fourth long-distance ligand, mainly phenylalanine and leucine. Distances and relationships among ligands and copper have been analyzed and compiled elsewhere [36] and represented in Table **1**. The structure of type-1 copper center in fungal laccases is trigonal planar and also in the laccase-like Fet3p protein, but appeared to have a distorted tetrahedral or trigonal pyramidal conformation in other multicopper oxidases from fungal or bacterial origin [88, 96].

In spite of earlier evidences, the global ligand environment of type-1 copper centers is not apparently related with the redox potential of the enzyme [8]. However, it has been recently suggested that the distance between the N atom of His 458 could be related with the increased redox potential of laccases like the *T. versicolor* one [81]. The same observation has been reported in the *R. lignosus* protein that showed a distorted conformation of the type-1 copper center [31].

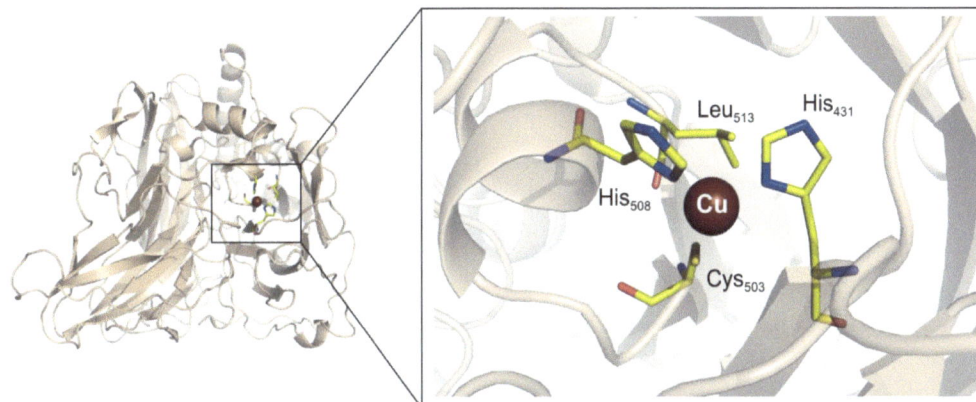

Figure 5: Location and structure of the coordination sphere for type-1 copper center from the *M. albomyces* laccase (PDB code: 1GW0) as obtained from X-ray crystallography studies. Copper coordination is ensured by a trigonal structure anchored in N and S atoms from Histidine and Cysteine residues, complemented by an axial ligand (Leucine).

Table 1: Aminoacids Involved in Copper Coordination in type-1 Copper Centers from Structurally Characterized Fungal Laccases[*]

PDB code	Source organism	First Ligand	Second Ligand	Third Ligand	Fourth ligand
1GW0	*Melanocarpus albomyces*	Cys 503 (2.20)	His 431 (1.91)	His 508 (1.93)	Leu 513 (3.68)
1GYC	*Trametes versicolor*	Cys 453 (2.19)	His 395 (2.02)	His 458 (2.04)	Phe 463 (3.65)
1KYA	*Coprinus cinereus*	Cys 452 (2.18)	His 396 (2.07)	His 457 (2.03)	Leu 462 (3.51)
1V10	*Rigidoporus lignosus*	Cys 452 (2.17)	His 396 (2.14)	His 457 (2.15)	Leu 462 (3.65)
2H5U	*Cerrena maxima*	Cys 453 (2.31)	His 395 (1.99)	His 458 (2.00)	Phe 463 (3.74)
2HRH	*Trametes trogii*	Cys 450 (2.14)	His 394 (2.09)	His 455 (2.09)	Phe 460 (3.74)
2HZH	*Coriolus zonatus*	Cys 453 (2.11)	His 395 (2.37)	His 458 (2.53)	Phe 463 (3.68)
2QT6	*Lentinus tigrinus*	Cys 452 (2.25)	His 394 (2.01)	His 457 (2.03)	Phe 462 (3.68)
2VDS	*Coriolopsis gallica*	Cys 449 (2.12)	His 393 (2.35)	His 454 (2.49)	Phe 459 (3.75)
3FPX	*Trametes hirsuta*	Cys 453 (2.24)	His 395 (2.01)	His 458 (2.05)	Phe 463 (3.63)

[*]Distances in Amgstrons between the coordinating atoms and the copper are represented in parentheses.

Trinuclear T2/T3 Copper Center

T2/T3 copper centers are characteristic of multicopper oxidases, and contain three copper atoms in two different conformations. Type-2 copper centers contain one single copper atom with a great variety of aminoacid ligands and coordination geometries. In these centers at least two of the coordinating aminoacids are histidines. The coordination sphere could be completed with different aminoacids such glutamate, glutamine or tyrosine [23, 64]. Absence of thiolate groups coordinating the copper atom result in a weak absortion band at 600 nm and the absence of blue color, characteristic of type-1 copper centers (see above). Type-3 centers are composed by two antiferromagnetically coupled copper atoms bridged by an oxygen species such as molecular oxygen or a hydroxyl group. They function as an electron storage mechanism, which allows laccases to transfer four electrons in a single step to reduce molecular oxygen.

Within laccases, these two copper centers (T2 and T3) are associated involving the coordination of eight histidine residues, belonging to the three cupredoxin domains of the protein. Only one histidine residue from the third terminal cupredoxin domain coordinates a copper atom in the T2/T3 center (Fig. **6**).

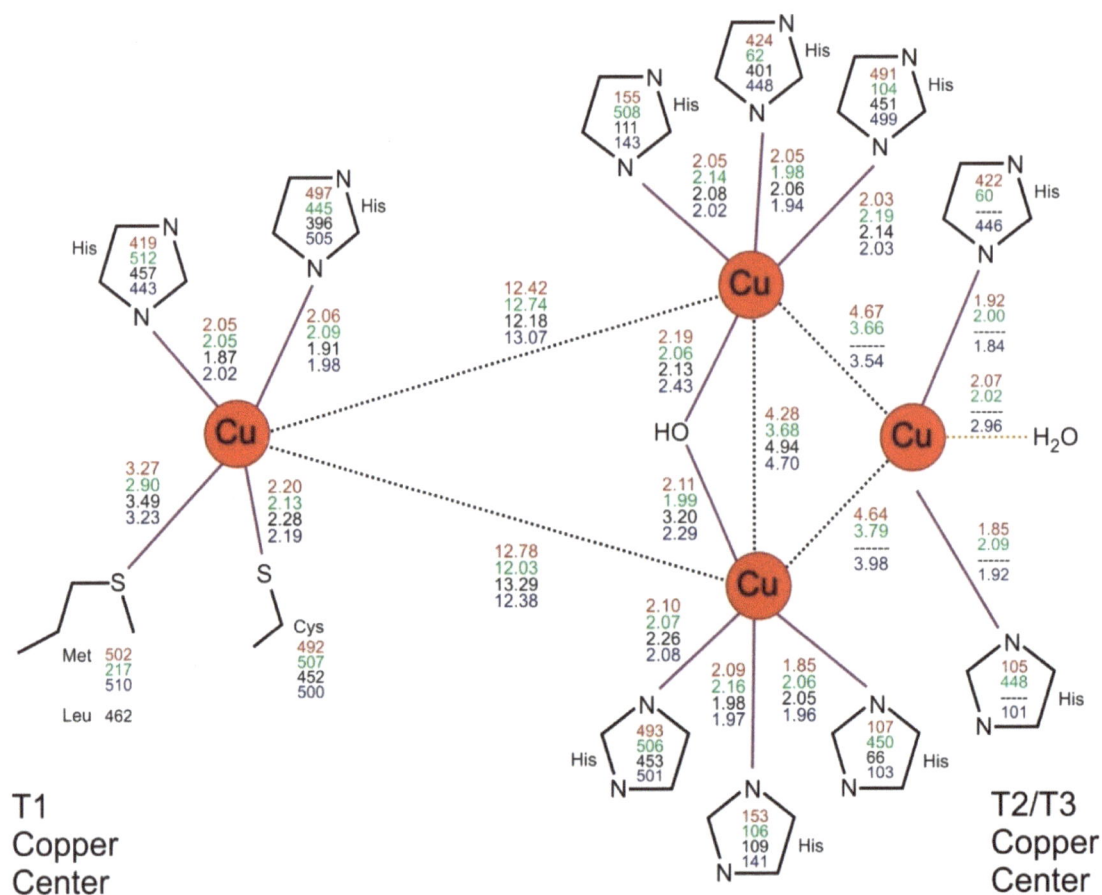

Figure 6: Overall arrangement of the reactive copper cluster in several representatives of the multicopper oxidase family. Residue number and distances between atoms are represented in colors: red, CotA laccase from *Bacillus subtilis* (PDB code: 1GSK); green, ascorbate oxidase from *Cucurbita pepo* (PDB code: 1AOZ); black, laccase from *C. cinereus* (PDB code: 1A65) and blue, CueO protein from *E. coli* (PDB code: 1KV7).

In fungal laccases the trinuclear copper center is structurally conserver among all the members of the family, showing a strong evolutive correlation between groups of specific fungi [10, 23, 24], but not with the prokaryotic relatives in which the fourth copper ligand is normally a methionine [27]. Interestingly, this evolutive conservation could be made extensive to other groups of laccases and more widely in multicopper oxidases (See Fig. 6). Fet3p protein from *S. cerevisiae* has been the major source of information about the molecular role of T2/T3 copper center in the enzymatic activity of fungal multicopper oxidases, that could be applied also to other members of the group [63, 79].

Studies performed in Fet3p protein using point mutations and specific inhibitors allowed to determine that electron transfer from T2/T3 copper center to an oxygen molecule is mediated by a peroxide intermediate in the same way that has been described in bacterial laccases [8]. However, the internal transfer and electron storage mechanisms that lead to the transport of electrons from the substrate to the Type-1 copper center and subsequently to the T2/T3 center and oxygen is still unclear.

Substrate Specificity, Protein Packing and Surface Properties

Laccases are classically defined by their ability to catalyze monoelectronic oxidations from a phenolic substrate different from tyrosine [5]. However, the unambiguous determination of laccase activity should be achieved by the isolation of the corresponding protein followed by determination of laccase activity against universal substrates like ABTS or catechol, compared with the activity over tyrosine for which laccase has little or no affinity [97].

A very wide range of substrates has been shown to be oxidized by the majority of fungal laccases with variable efficiency depending on the source microorganism [50, 54, 97]. Higher Km values have been determined for substrates like ABTS and syringaldazine (within the range from 1-10 μM). Some laccases like the one isolated from *T. versicolor*, have been reported as able to oxidize with a low efficiency very complex phenolic compounds from the family of polycyclic aromatic hydrocarbons (PAHs) and chloro-phenols [50, 59, 66]. A comprehensive review on the fungal laccase substrates has been previously published by Baldrian [5].

However, until only a few examples of structure determinations of laccases adducts with their substrates have been reported so far. These include the *T. versicolor* protein complexed with 2,5-xylidine [10], the *M. albomyces* laccase in complex with 2,6-methoxyphenol and the prokaryotic CotA laccase from *B. subtilis* in complex with ABTS [26]. (Figs. **7** and **8**).

Structures of laccase-substrate complexes showed the presence of substrate binding cavities with different sizes in all the members of the family. CotA laccase substrate pocket is more open than the cavities present in fungal laccases (Figs. **7** and **8**). However, the electrostatic properties of the substrate binding cavities are rather similar in all the already determined complexes. Very interesting observations pointed in the direction that also the lipophilicity potential of the active site of fungal and bacterial laccases showed a characteristic pattern, dominated mainly by hydrophilic patches close to the substrate binding site (Fig. **8**). In the particular case of the *M. albomyces* laccase, the structural information obtained from the adduct complex has been used to propose an electron transfer pathway from the substrate to the Type-1 copper center [45], and also to derive potential residues for targeted engineering of the enzyme to improve its substrate range [54].

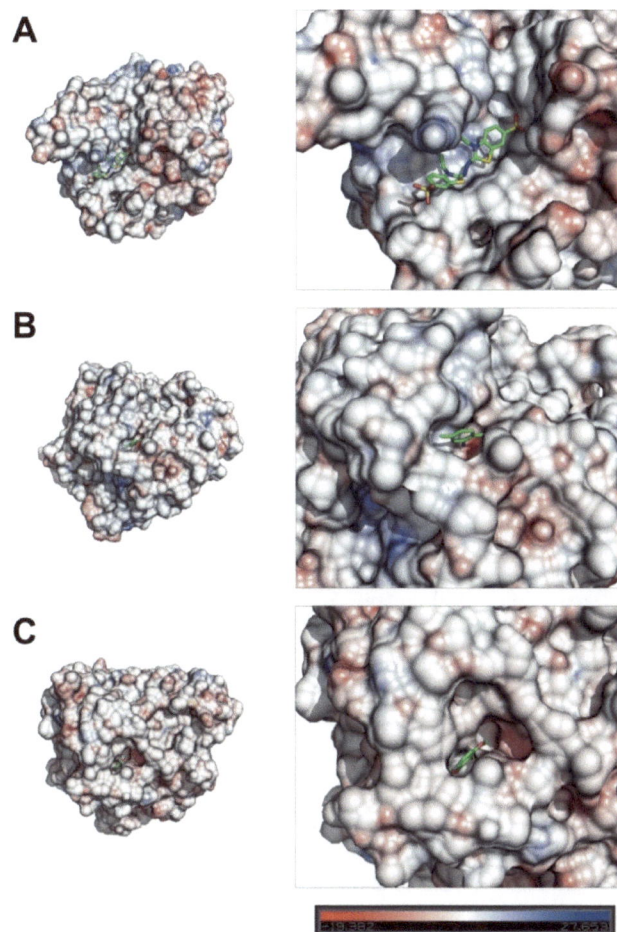

Figure 7: Electrostatic potential of selected laccases complexed with their substrates showing a detailed view of the substrate binding pocket. Panel A, CotA bacterial laccase from *B. subtilis* complexed with ABTS (PDB code: 1UVW); Panel B, laccase from *T. versicolor* complexed with 2,5-xylidine (PDB code: 1KYA) and Panel C, laccase from *M. albomyces* complexed with 2,5-dimethoxyphenol (PDB code: 3FU7). Electrostatic potential calculations were performed using VASCo software [101] and the representations prepared with the Pymol molecular graphics system [21].

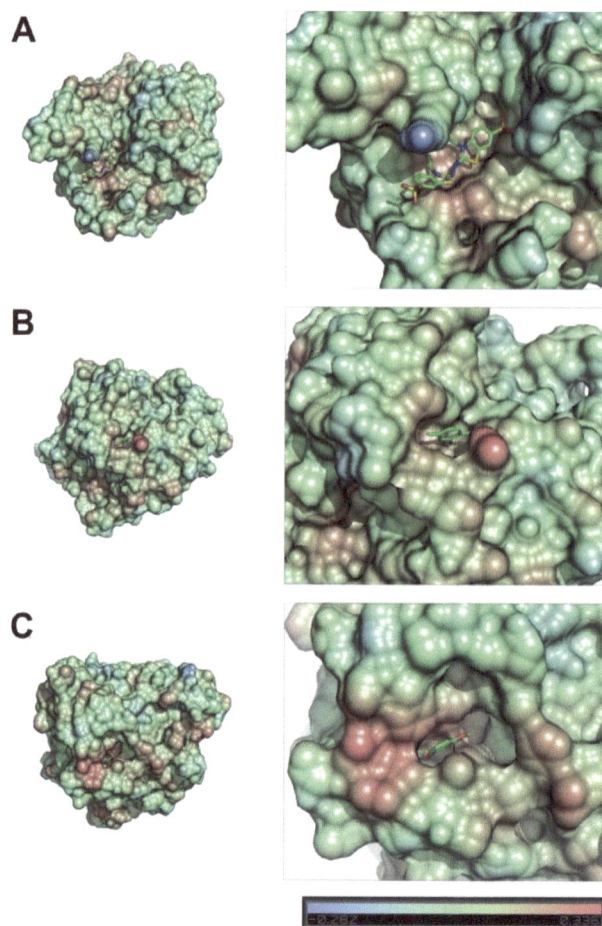

Figure 8: Molecular lipophilicity potential (MLP) of selected laccases complexed with their substrates showing a detailed view of the substrate binding pocket. Panel A, CotA bacterial laccase from *B. subtilis* complexed with ABTS (PDB code: 1UVW); Panel B, laccase from *T. versicolor* complexed with 2,5-xylidine (PDB code: 1KYA) and Panel C, laccase from *M. albomyces* complexed with 2,5-dimethoxyphenol (PDB code: 3FU7). MLP calculations were performed using VASCo software [101] and the panels prepared with the Pymol molecular graphics system [21].

TYROSINASES

Introduction

Tyrosinases are copper proteins that belong to a wider group of enzymes including also the catecholoxidases from plants and the haemocyanins (oxigen-carrier proteins) from arthropods [39, 84]. Tyrosinases are type-III copper enzymes involved in the melanin biosynthetic pathway, specifically in the first step of the melanin production (conversion of L-tyrosine into L-dopaquinone). Enzymatically, tyrosinases are phenol hydroxilases able to catalyze an o-hydroxylation of monophenols, and the subsequent oxidation of these compounds to quinones, using molecular oxygen. The originated quinonic species will undergo spontaneous non-enzymatic reactions to produce complex phenolic polymers with a characteristic dark brown colour [39, 84, 93].

Tyrosinases are present in mammals, invertebrates, plants and microorganisms [84, 93]. These enzymes were first studied and characterized in mammals, because of their central role in the development of melanomas and diseases related with pigmentation [4]. Tyrosinases are essential enzymes in melanin biosynthesis and therefore responsible for pigmentation of skin and hair in mammals, where two more enzymes, the tyrosinase-related proteins (Tyrps), participate in the pathway [78]. However, the detailed biochemical characterization of tyrosinases was accomplished in consequence of its main role in the production of a melanin-line brownish pigment by the mushroom *Agaricus bisporus* that prevent its preservation for prolonged periods and its commercial use in food industry [11, 30, 111]. In fungi, tyrosinases are cytosolic enzymes, which showed an increased variability in sequence if compared with other groups of multicopper oxidases such as laccases [39, 57]. In all organisms, tyrosinases exist in two forms; one latent and other active [93]. In fungi, latent form of tyrosinases have been shown to be *in vitro* activated mainly by environmental changes such as pH and chemical stress [87]. The mechanism for the fungal tyrosinase activation is still unclear, however it has been related with protein folding phenomena and also with a proteolytic processintg by the help of cytoplasmic proteases [32, 40, 49, 65].

Phylogenetic analysis based on tyrosinase sequences showed that fungal tyrosinases can be clustered in groups for basidiomycetes, deuteromycetes and ascomycetes (Fig. **9**). Tyrosinase sequences are not very conserved, in despite of the aminoacids involved in copper coordination. This heterogeneity is also observable at the gene level concerning gene size, exon number and sequence homology among the group [39, 49, 92].

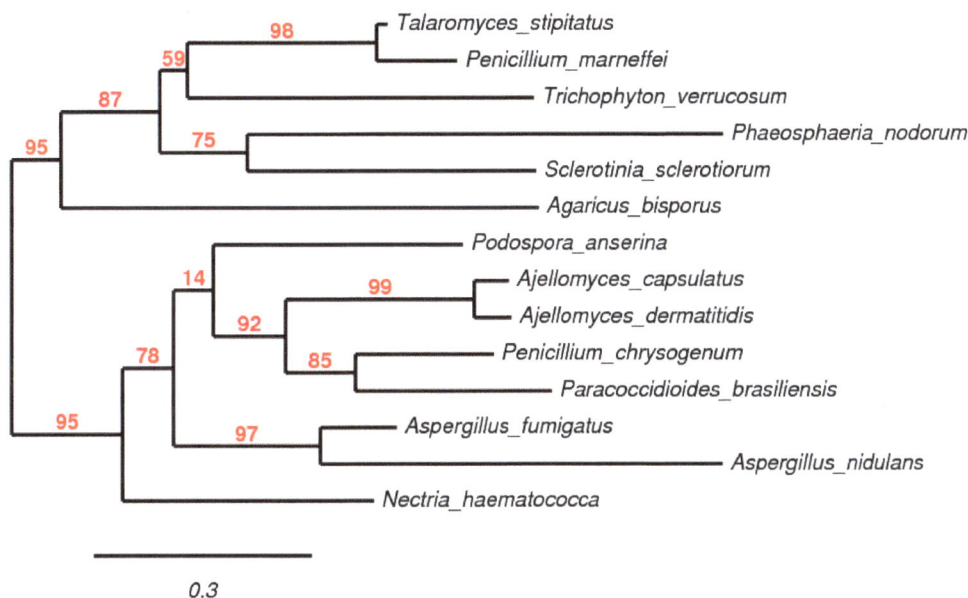

Figure 9: Phylogenetic tree obtained from the sequence alignment of fungal tyrosinases by the MULTALIN algorythm [12]. Distances between nodes are represented by numbers. Tree was prepared with Bosque software [83].

Copper Center and Catalytic Properties

Tyrosinases were one of the first monooxygenases discovered and biochemically characterized in their reaction mechanism. However, there is only one crystallographic structure available for a tyrosinase from the filamentous bacterium *Streptomyces castaneoglobisporus* (PDB code: 2ZWE) [67].

Haemocyanins, catechol oxidases and tyrosinases share the same binuclear copper center, in which two anti-ferromagnetically coupled copper atoms are coordinated exclusively by six histidine residues (See Fig. **10** for a detailed structural representation). Fungal tyrosinases showed a high sequence homology with the already characterized tyrosinase from *S. castaneoglobisporus* in the region containing the histidine residues involved in copper coordination [39, 93].

The original motivation for the study of tyrosinases from fungi (mainly in food applications) was the browning phenomenon during harvesting, processing and storage. In this area, a lot of effort has been made to develop tyrosinase inhibitors to be used either in food or in pharmaceutical industry [15, 51, 78, 89]. However, fungal tyrosinases constitute a wider research topic in which the most important goals are related with their catalytic properties and potential biotechnological applications [40, 41, 49, 87].

Together with the classical biotechnological applications in the degradation of phenolic compounds, one of the most interesting characteristics of tyrosinases is their ability to catalyze covalent crosslinkings between phenol monomers to produce diphenols that can be used as antioxidant agents in food additives or pharmaceutical formulations [73, 78,

84]. Polymerization of phenols in a controlled manner could be also a potential application of fungal tyrosinases [39, 41]. Recently, the properties of the catalytic centers of tyrosinases have been studied by several approaches, including molecular modelling [20, 78]. Catalytic center of tyrosinases showed in general a more open cavity than the observed in laccases, facilitating the access to a potential substrate of the enzyme. Catalytic copper atoms from tyrosinases are located very close to the surface of the protein, in clear contrast to the properties of the laccase active centers [26]. These characteristics are probably also related with the wide range of possible substrates already characterized for the fungal and bacterial tyrosinases [28, 41, 44, 72, 89].

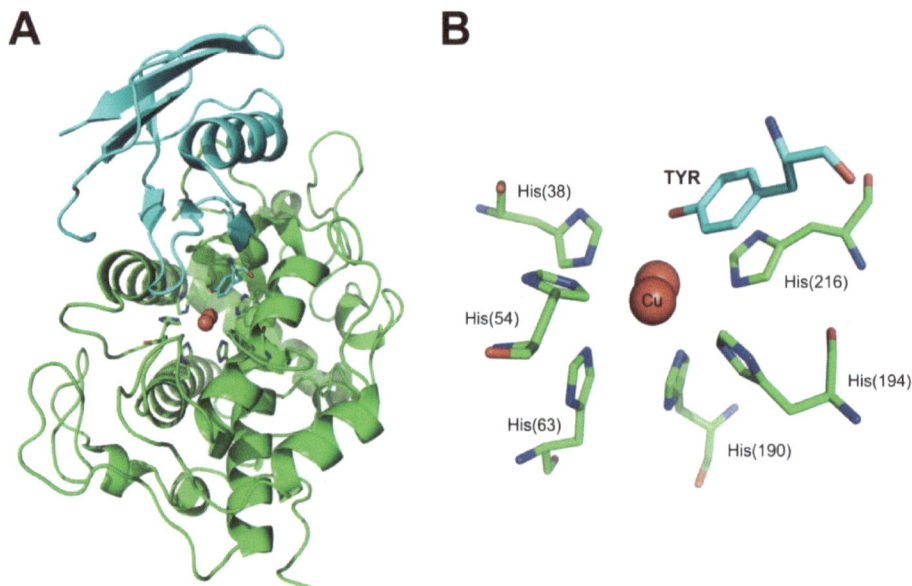

Figure 10: Structural features of tyrosinase from *Streptomyces castaneoglobisporus*. Panel A, overall tertiary structure of the copper-bound tyrosinase (represented in green) complexed with a caddie protein (represented in blue) as solved by X-ray crystallography (PDB code: 2ZWE) [67]. Copper atoms are represented as red balls and the residues involved in copper coordination are shown in sticks. Panel B, detailed representation of the coordination sphere of the binuclear copper center and position of the tyrosine substrate. Figure was prepared with Pymol [21].

Fungal tyrosinases showed clear biochemical advantages in their possible applications over laccases because of their lower molecular weight, wider substrate range and the presence of a less complex catalytic core (two copper atoms instead of four) [20, 39, 78]. However they are generally intracellular and produced in a low amounts if we compare them with other fungal oxidases such as laccases, and their stability is generally lower at physiological conditions. Homologous and heterologous host expression of fungal tyrosinase genes will allow in a near future to know more in detail the biochemical properties of these enzymes, and to overcome their limitations in order to improve their stability and expand their applications in the food and pharmaceutical biotechnology [40, 89, 108].

CONCLUSION AND FURTHER PERSPECTIVES

The unique combination of the chemical properties of copper atoms together with the biological help of polypeptides constitutes a widespread system to catalyze biological oxidations over a great variety of substrates. Among the family of multicopper oxidases, we have analyzed the structural and biochemical properties of two main groups: laccases and tyrosinases. Structural biology contribution has been essential for the understanding of the role and physiology of these enzymes, but also for their engineering and design in future applications. Knowledge of the catalytic and structural properties of both of the protein families are currently being used for the improvement of the application range for these enzymes. These facts are specifically true for both of the analyzed families of enzymes. Multicopper oxidases are starting to be main players on biotechnological applications involving oxidation of aromatic compounds, such as bioremediation of pollutants, biotransformations of leader compounds, organic semi-synthesis and others. More efforts in the direction of the use of recombinant DNA technology will help to the production and design of more stable and more efficient multicopper oxidases for their use in a wider range of applications.

REFERENCES

[1] Abraham ZH, Smith BE, Howes BD, *et al.* pH-dependence for binding a single nitrite ion to each type-2 copper centre in the copper-containing nitrite reductase of *Alcaligenes xylosoxidans.* Biochem J 1997; 324 (Pt 2): 511-6.

[2] Ahmed S, Riaz S, Jamil A. Molecular cloning of fungal xylanases: an overview. Appl Microbiol Biotechnol 2009; 84: 19-35.

[3] Ayala M, Pickard MA, Vazquez-Duhalt R. Fungal enzymes for environmental purposes, a molecular biology challenge. J Mol Microbiol Biotechnol 2008; 15: 172-80.

[4] Baldea I, Mocan T, Cosgarea R. The role of ultraviolet radiation and tyrosine stimulated melanogenesis in the induction of oxidative stress alterations in fair skin melanocytes. Exp Oncol 2009; 31: 200-8.

[5] Baldrian P. Fungal laccases - occurrence and properties. FEMS Microbiol Rev 2006; 30: 215-42.

[6] Battistuzzi G, Bellei M, Leonardi A, *et al.* Reduction thermodynamics of the T1 Cu site in plant and fungal laccases. J Biol Inorg Chem 2005; 10: 867-73.

[7] Ben-Ami R, Lewis RE, Leventakos K, Kontoyiannis DP. *Aspergillus fumigatus* inhibits angiogenesis through the production of gliotoxin and other secondary metabolites. Blood 2009; 114: 5393-9.

[8] Bento I, Martins LO, Gato Lopes G, *et al.* Dioxygen reduction by multi-copper oxidases; a structural perspective. Dalton Trans 2005; 3507-13.

[9] Berry SM, Mayers JR, Zehm NA. Models of noncoupled dinuclear copper centers in azurin. J Biol Inorg Chem 2009; 14: 143-9.

[10] Bertrand T, Jolivalt C, Briozzo P, *et al.* Crystal structure of a four-copper laccase complexed with an arylamine: insights into substrate recognition and correlation with kinetics. Biochemistry 2002; 41: 7325-33.

[11] Boekelheide K, Graham DG, Mize PD, Jeffs PW. The metabolic pathway catalyzed by the tyrosinase of *Agaricus bisporus.* J Biol Chem 1980; 255: 4766-71.

[12] Candresse T, Morch MD, Dunez J. Multiple alignment and hierarchical clustering of conserved amino acid sequences in the replication-associated proteins of plant RNA viruses. Res Virol 1990; 141: 315-29.

[13] Chang A, Scheer M, Grote A, *et al.* BRENDA, AMENDA and FRENDA the enzyme information system: new content and tools in 2009. Nucleic Acids Res 2009; 37: D588-92.

[14] Chemier JA, Fowler ZL, Koffas MA, Leonard E. Trends in microbial synthesis of natural products and biofuels. Adv Enzymol Relat Areas Mol Biol 2009; 76: 151-217.

[15] Chiku K, Dohi H, Saito A, *et al.* Enzymatic synthesis of 4-hydroxyphenyl beta-D-oligoxylosides and their notable tyrosinase inhibitory activity. Biosci Biotechnol Biochem 2009; 73: 1123-8.

[16] Claus H. Laccases and their occurrence in prokaryotes. Arch Microbiol 2003; 179: 145-50.

[17] Claus H. Laccases: structure, reactions, distribution. Micron 2004; 35: 93-6.

[18] Crichton RR, Pierre JL. Old iron, young copper: from Mars to Venus. Biometals 2001; 14: 99-112.

[19] de Silva D, Davis-Kaplan S, Fergestad J, Kaplan J. Purification and characterization of Fet3 protein, a yeast homologue of ceruloplasmin. J Biol Chem 1997; 272: 14208-13.

[20] Deeth RJ, Diedrich C. Structural and mechanistic insights into the oxy form of tyrosinase from molecular dynamics simulations. J Biol Inorg Chem 2010; 15: 117-29.

[21] DeLano WL. The PyMOL Molecular Graphics System. DeLano Scientific, San Carlos, CA, USA 2002.

[22] Do BC, Dang TT, Berrin JG, *et al.* Cloning, expression in *Pichia pastoris,* and characterization of a thermostable GH5 mannan endo-1,4-beta-mannosidase from *Aspergillus niger* BK01. Microb Cell Fact 2009; 8: 59.

[23] Ducros V, Brzozowski AM, Wilson KS, *et al.* Crystal structure of the type-2 Cu depleted laccase from *Coprinus cinereus* at 2.2 A resolution. Nat Struct Biol 1998; 5: 310-6.

[24] Ducros V, Brzozowski AM, Wilson KS, *et al.* Structure of the laccase from *Coprinus cinereus* at 1.68 A resolution: evidence for different 'type 2 Cu-depleted' isoforms. Acta Crystallogr D Biol Crystallogr 2001; 57: 333-6.

[25] Dutta T, Sahoo R, Sengupta R, *et al.* Novel cellulases from an extremophilic filamentous fungi *Penicillium citrinum*: production and characterization. J Ind Microbiol Biotechnol 2008; 35: 275-82.

[26] Enguita FJ, Marcal D, Martins LO, *et al.* Substrate and dioxygen binding to the endospore coat laccase from *Bacillus subtilis.* J Biol Chem 2004; 279: 23472-6.

[27] Enguita FJ, Martins LO, Henriques AO, Carrondo MA. Crystal structure of a bacterial endospore coat component. A laccase with enhanced thermostability properties. J Biol Chem 2003; 278: 19416-25.

[28] Fernandez E, Sanchez-Amat A, Solano F. Location and catalytic characteristics of a multipotent bacterial polyphenol oxidase. Pigment Cell Res 1999; 12: 331-9.

[29] Ferraroni M, Myasoedova NM, Schmatchenko V, *et al.* Crystal structure of a blue laccase from *Lentinus tigrinus*: evidences for intermediates in the molecular oxygen reductive splitting by multicopper oxidases. BMC Struct Biol 2007; 7: 60.

[30]   Fry DC, Strothkamp KG. Photoinactivation of *Agaricus bisporus* tyrosinase: modification of the binuclear copper site. Biochemistry 1983; 22: 4949-53.

[31]   Garavaglia S, Cambria MT, Miglio M, *et al.* The structure of *Rigidoporus lignosus* Laccase containing a full complement of copper ions, reveals an asymmetrical arrangement for the T3 copper pair. J Mol Biol 2004; 342: 1519-31.

[32]   Garcia-Moreno M, Rodriguez-Lopez JN, Martinez-Ortiz F, *et al.* Effect of pH on the oxidation pathway of dopamine catalyzed by tyrosinase. Arch Biochem Biophys 1991; 288: 427-34.

[33]   Gardiner DM, Kazan K, Manners JM. Novel genes of *Fusarium graminearum* that negatively regulate deoxynivalenol production and virulence. Mol Plant Microbe Interact 2009; 22: 1588-600.

[34]   Georgianna DR, Payne GA. Genetic regulation of aflatoxin biosynthesis: from gene to genome. Fungal Genet Biol 2009; 46: 113-25.

[35]   Gouet P, Robert X, Courcelle E. ESPript/ENDscript: Extracting and rendering sequence and 3D information from atomic structures of proteins. Nucleic Acids Res 2003; 31: 3320-3.

[36]   Gray HB, Malmstrom BG, Williams RJ. Copper coordination in blue proteins. J Biol Inorg Chem 2000; 5: 551-9.

[37]   Gromov I, Marchesini A, Farver O, *et al.* Azide binding to the trinuclear copper center in laccase and ascorbate oxidase. Eur J Biochem 1999; 266: 820-30.

[38]   Hakulinen N, Andberg M, Kallio J, *et al.* A near atomic resolution structure of a *Melanocarpus albomyces* laccase. J Struct Biol 2008; 162: 29-39.

[39]   Halaouli S, Asther M, Sigoillot JC, *et al.* Fungal tyrosinases: new prospects in molecular characteristics, bioengineering and biotechnological applications. J Appl Microbiol 2006; 100: 219-32.

[40]   Halaouli S, Record E, Casalot L, *et al.* Cloning and characterization of a tyrosinase gene from the white-rot fungus *Pycnoporus sanguineus*, and overproduction of the recombinant protein in *Aspergillus niger*. Appl Microbiol Biotechnol 2006; 70: 580-9.

[41]   Hernandez-Romero D, Solano F, Sanchez-Amat A. Polyphenol oxidase activity expression in *Ralstonia solanacearum*. Appl Environ Microbiol 2005; 71: 6808-15.

[42]   Hofer C, Schlosser D. Novel enzymatic oxidation of Mn^{2+} to Mn^{3+} catalyzed by a fungal laccase. FEBS Lett 1999; 451: 186-90.

[43]   Ji HF, Zhang HY. Bioinformatic identification of the most ancient copper protein architecture. J Biomol Struct Dyn 2008; 26: 197-201.

[44]   Kahng HY, Chung BS, Lee DH, *et al. Cellulophaga tyrosinoxydans* sp. nov., a tyrosinase-producing bacterium isolated from seawater. Int J Syst Evol Microbiol 2009; 59: 654-7.

[45]   Kallio JP, Auer S, Janis J, *et al.* Structure-Function Studies of a *Melanocarpus albomyces* Laccase Suggest a Pathway for Oxidation of Phenolic Compounds. J Mol Biol 2009;

[46]   Kaneko S, Cheng M, Murai H, *et al.* Purification and characterization of an extracellular laccase from *Phlebia radiata* strain BP-11-2 that decolorizes fungal melanin. Biosci Biotechnol Biochem 2009; 73: 939-42.

[47]   Kataoka K, Sugiyama R, Hirota S, *et al.* Four-electron reduction of dioxygen by a multicopper oxidase, CueO, and roles of Asp112 and Glu506 located adjacent to the trinuclear copper center. J Biol Chem 2009; 284: 14405-13.

[48]   Kataoka K, Tsukamoto K, Kitagawa R, *et al.* Compensatory binding of an asparagine residue to the coordination-unsaturated type I Cu center in bilirubin oxidase mutants. Biochem Biophys Res Commun 2008; 371: 416-9.

[49]   Kawamura-Konishi Y, Tsuji M, Hatana S, *et al.* Purification, characterization, and molecular cloning of tyrosinase from *Pholiota nameko*. Biosci Biotechnol Biochem 2007; 71: 1752-60.

[50]   Keum YS, Li QX. Fungal laccase-catalyzed degradation of hydroxy polychlorinated biphenyls. Chemosphere 2004; 56: 23-30.

[51]   Kim YJ, Uyama H. Tyrosinase inhibitors from natural and synthetic sources: structure, inhibition mechanism and perspective for the future. Cell Mol Life Sci 2005; 62: 1707-23.

[52]   Kobayashi M, Shimizu S. Cobalt proteins. Eur J Biochem 1999; 261: 1-9.

[53]   Kobayashi M, Shoun H. The copper-containing dissimilatory nitrite reductase involved in the denitrifying system of the fungus *Fusarium oxysporum*. J Biol Chem 1995; 270: 4146-51.

[54]   Kunamneni A, Camarero S, Garcia-Burgos C, *et al.* Engineering and Applications of fungal laccases for organic synthesis. Microb Cell Fact 2008; 7: 32.

[55]   Kwok EY, Severance S, Kosman DJ. Evidence for iron channeling in the Fet3p-Ftr1p high-affinity iron uptake complex in the yeast plasma membrane. Biochemistry 2006; 45: 6317-27.

[56]   Larrondo LF, Salas L, Melo F, *et al.* A novel extracellular multicopper oxidase from *Phanerochaete chrysosporium* with ferroxidase activity. Appl Environ Microbiol 2003; 69: 6257-63.

[57]   Laufer Z, Beckett RP, Minibayeva FV. Co-occurrence of the multicopper oxidases tyrosinase and laccase in lichens in sub-order peltigerineae. Ann Bot (Lond) 2006; 98: 1035-42.

[58]   Leonowicz A, Cho NS, Luterek J, et al. Fungal laccase: properties and activity on lignin. J Basic Microbiol 2001; 41: 185-227.

[59]   Leontievsky AA, Myasoedova NM, Baskunov BP, *et al.* Transformation of 2,4,6-trichlorophenol by free and immobilized fungal laccase. Appl Microbiol Biotechnol 2001; 57: 85-91.

[60]   Leontievsky AA, Vares T, Lankinen P, *et al.* Blue and yellow laccases of ligninolytic fungi. FEMS Microbiol Lett 1997; 156: 9-14.

[61]   Ludwig ML, Matthews RG. Structure-based perspectives on B12-dependent enzymes. Annu Rev Biochem 1997; 66: 269-313.

[62]   Lyashenko AV, Bento I, Zaitsev VN, *et al.* X-ray structural studies of the fungal laccase from *Cerrena maxima*. J Biol Inorg Chem 2006; 11: 963-73.

[63]   Machonkin TE, Quintanar L, Palmer AE, et al. Spectroscopy and reactivity of the type 1 copper site in Fet3p from *Saccharomyces cerevisiae*: correlation of structure with reactivity in the multicopper oxidases. J Am Chem Soc 2001; 123: 5507-17.

[64]   MacPherson IS, Murphy ME. Type-2 copper-containing enzymes. Cell Mol Life Sci 2007; 64: 2887-99.

[65]   Madani W, Kermasha S, Versari A. Characterization of tyrosinase- and polyphenol esterase-catalyzed end products using selected phenolic substrates. J Agric Food Chem 1999; 47: 2486-90.

[66]   Majcherczyk A, Johannes C. Radical mediated indirect oxidation of a PEG-coupled polycyclic aromatic hydrocarbon (PAH) model compound by fungal laccase. Biochim Biophys Acta 2000; 1474: 157-62.

[67]   Matoba Y, Kumagai T, Yamamoto A, *et al.* Crystallographic evidence that the dinuclear copper center of tyrosinase is flexible during catalysis. J Biol Chem 2006; 281: 8981-90.

[68]   Mendel RR, Bittner F. Cell biology of molybdenum. Biochim Biophys Acta 2006; 1763: 621-35.

[69]   Mikolasch A, Schauer F. Fungal laccases as tools for the synthesis of new hybrid molecules and biomaterials. Appl Microbiol Biotechnol 2009; 82: 605-24.

[70]   Morozova OV, Shumakovich GP, Gorbacheva MA, *et al.* "Blue" laccases. Biochemistry (Mosc) 2007; 72: 1136-50.

[71]   Moura JJ, Brondino CD, Trincao J, Romao MJ. Mo and W bis-MGD enzymes: nitrate reductases and formate dehydrogenases. J Biol Inorg Chem 2004; 9: 791-9.

[72]   Munoz-Munoz JL, Garcia-Molina F, Garcia-Molina M, *et al.* Ellagic acid: characterization as substrate of polyphenol oxidase. IUBMB Life 2009; 61: 171-7.

[73]   Munoz-Munoz JL, Garcia-Molina F, Varon R, *et al.* Kinetic characterization of the oxidation of esculetin by polyphenol oxidase and peroxidase. Biosci Biotechnol Biochem 2007; 71: 390-6.

[74]   Murphy ME, Lindley PF, Adman ET. Structural comparison of cupredoxin domains: domain recycling to construct proteins with novel functions. Protein Sci 1997; 6: 761-70.

[75]   Nagai M, Kawata M, Watanabe H, *et al.* Important role of fungal intracellular laccase for melanin synthesis: purification and characterization of an intracellular laccase from *Lentinula edodes* fruit bodies. Microbiology 2003; 149: 2455-62.

[76]   Nakamura K, Go N. Function and molecular evolution of multicopper blue proteins. Cell Mol Life Sci 2005; 62: 2050-66.

[77]   Olesen K, Veselov A, Zhao Y, *et al.* Spectroscopic, kinetic, and electrochemical characterization of heterologously expressed wild-type and mutant forms of copper-containing nitrite reductase from *Rhodobacter sphaeroides* 2.4.3. Biochemistry 1998; 37: 6086-94.

[78]   Olivares C, Solano F. New insights into the active site structure and catalytic mechanism of tyrosinase and its related proteins. Pigment Cell Melanoma Res 2009; 22: 750-60.

[79]   Palmer AE, Quintanar L, Severance S, *et al.* Spectroscopic characterization and O_2 reactivity of the trinuclear Cu cluster of mutants of the multicopper oxidase Fet3p. Biochemistry 2002; 41: 6438-48.

[80]   Pinho D, Besson S, Brondino CD, *et al.* Copper-containing nitrite reductase from *Pseudomonas chlororaphis* DSM 50135. Eur J Biochem 2004; 271: 2361-9.

[81]   Piontek K, Antorini M, Choinowski T. Crystal structure of a laccase from the fungus *Trametes versicolor* at 1.90-A resolution containing a full complement of coppers. J Biol Chem 2002; 277: 37663-9.

[82]   Prudencio M, Eady RR, Sawers G. The blue copper-containing nitrite reductase from *Alcaligenes xylosoxidans:* cloning of the nirA gene and characterization of the recombinant enzyme. J Bacteriol 1999; 181: 2323-9.

[83]   Ramirez-Flandes S, Ulloa O. Bosque: integrated phylogenetic analysis software. Bioinformatics 2008; 24: 2539-41.

[84]   Rast DM, Baumgartner D, Mayer C, Hollenstein GO. Cell wall-associated enzymes in fungi. Phytochemistry 2003; 64: 339-66.

[85]   Rees DC. Great metalloclusters in enzymology. Annu Rev Biochem 2002; 71: 221-46.

[86]   Riva S. Laccases: blue enzymes for green chemistry. Trends Biotechnol 2006; 24: 219-26.

[87]   Sakamoto Y, Nakade K, Sato T. Characterization of the post-harvest changes in gene transcription in the gill of the Lentinula edodes fruiting body. Curr Genet 2009; 55: 409-23.

[88]   Sakurai T, Kataoka K. Structure and function of type I copper in multicopper oxidases. Cell Mol Life Sci 2007; 64: 2642-56.

[89]   Sanchez-Amat A, Lucas-Elio P, Fernandez E, *et al.* Molecular cloning and functional characterization of a unique multipotent polyphenol oxidase from *Marinomonas mediterranea*. Biochim Biophys Acta 2001; 1547: 104-16.

[90]   Schmidt-Heydt M, Magan N, Geisen R. Stress induction of mycotoxin biosynthesis genes by abiotic factors. FEMS Microbiol Lett 2008; 284: 142-9.

[91]   Sedlak E, Ziegler L, Kosman DJ, Wittung-Stafshede P. *In vitro* unfolding of yeast multicopper oxidase Fet3p variants reveals unique role of each metal site. Proc Natl Acad Sci U S A 2008; 105: 19258-63.

[92]   Selinheimo E, Saloheimo M, Ahola E, *et al.* Production and characterization of a secreted, C-terminally processed tyrosinase from the filamentous fungus *Trichoderma reesei*. Febs J 2006; 273: 4322-35.

[93]   Seo SY, Sharma VK, Sharma N. Mushroom tyrosinase: recent prospects. J Agric Food Chem 2003; 51: 2837-53.

[94]   Shatsky M, Nussinov R, Wolfson HJ. A method for simultaneous alignment of multiple protein structures. Proteins 2004; 56: 143-56.

[95]   Shephard GS. Impact of mycotoxins on human health in developing countries. Food Addit Contam Part A Chem Anal Control Expo Risk Assess 2008; 25: 146-51.

[96]   Shi X, Stoj C, Romeo A, *et al.* Fre1p Cu^{+2} reduction and Fet3p Cu^{+3} oxidation modulate copper toxicity in *Saccharomyces cerevisiae*. J Biol Chem 2003; 278: 50309-15.

[97]   Shumakovich GP, Shleev SV, Morozova OV, *et al.* Electrochemistry and kinetics of fungal laccase mediators. Bioelectrochemistry 2006; 69: 16-24.

[98]   Shwab EK, Keller NP. Regulation of secondary metabolite production in filamentous ascomycetes. Mycol Res 2008; 112: 225-30.

[99]   Sigoillot C, Record E, Belle V, *et al.* Natural and recombinant fungal laccases for paper pulp bleaching. Appl Microbiol Biotechnol 2004; 64: 346-52.

[100]   Singh Arora D, Kumar Sharma R. Ligninolytic Fungal Laccases and Their Biotechnological Applications. Appl Biochem Biotechnol 2010; 160: 1760-88.

[101]   Steinkellner G, Rader R, Thallinger GG, *et al.* VASCo: computation and visualization of annotated protein surface contacts. BMC Bioinformatics 2009; 10: 32.

[102]   Stepanova EV, Fedorova TV, Sorokina ON, *et al.* Effect of solvent phase transitions on enzymatic activity and structure of laccase from *Coriolus hirsutus*. Biochemistry (Mosc) 2009; 74: 385-92.

[103]   Stoj C, Kosman DJ. Cuprous oxidase activity of yeast Fet3p and human ceruloplasmin: implication for function. FEBS Lett 2003; 554: 422-6.

[104]   Susla M, Novotny C, Svobodova K. The implication of *Dichomitus squalens* laccase isoenzymes in dye decolorization by immobilized fungal cultures. Bioresour Technol 2007; 98: 2109-15.

[105]   Vandenberghe IH, Meyer TE, Cusanovich MA, Van Beeumen JJ. The covalent structure of the blue copper-containing nitrite reductase from *Achromobacter xylosoxidans*. Biochem Biophys Res Commun 1998; 247: 734-40.

[106]   Veeramalai M, Ye Y, Godzik A. TOPS++FATCAT: fast flexible structural alignment using constraints derived from TOPS+ Strings Model. BMC Bioinformatics 2008; 9: 358.

[107]   Veselov A, Olesen K, Sienkiewicz A, *et al.* Electronic structural information from Q-band ENDOR on the type 1 and type 2 copper liganding environment in wild-type and mutant forms of copper-containing nitrite reductase. Biochemistry 1998; 37: 6095-105.

[108]   Westerholm-Parvinen A, Selinheimo E, Boer H, *et al.* Expression of the *Trichoderma reesei* tyrosinase 2 in *Pichia pastoris*: isotopic labeling and physicochemical characterization. Protein Expr Purif 2007; 55: 147-58.

[109]   Wijma HJ, Canters GW, de Vries S, Verbeet MP. Bidirectional catalysis by copper-containing nitrite reductase. Biochemistry 2004; 43: 10467-74.

[110]   Wong DW. Structure and action mechanism of ligninolytic enzymes. Appl Biochem Biotechnol 2009; 157: 174-209.

[111]   Yamaguchi M, Hwang PM, Campbell JD. Latent o-diphenol oxidase in mushrooms (*Agaricus bisporus*). Can J Biochem 1970; 48: 198-202.

[112]   Ye Y, Godzik A. FATCAT: a web server for flexible structure comparison and structure similarity searching. Nucleic Acids Res 2004; 32: W582-5.

[113]   Yoshida H. Chemistry of lacquer (Urushi), Part 1. J Chem Soc 1883; 43: 472-86.

[114]   Yu J, Payne GA, Nierman WC, *et al. Aspergillus flavus* genomics as a tool for studying the mechanism of aflatoxin formation. Food Addit Contam Part A Chem Anal Control Expo Risk Assess 2008; 25: 1152-7.

[115]   Zhong JJ, Xiao JH. Secondary metabolites from higher fungi: discovery, bioactivity, and bioproduction. Adv Biochem Eng Biotechnol 2009; 113: 79-150.

CHAPTER 5

Mold-Fermented Foods: *Penicillium* spp. as Ripening Agents in the Elaboration of Cheese and Meat Products

Renato Chávez[1], Francisco Fierro[2,*], Ramón O. García-Rico[3] and Federico Laich[4]

[1]Departamento de Biología, Facultad de Química y Biología, Universidad de Santiago de Chile (USACH), Santiago, Chile. [2]Departamento de Biotecnología, División de Ciencias Biológicas y de la Salud, Universidad Autonóma Metropolitana-Iztapalapa, México D.F. [3]Departamento de Microbiología, Facultad de Ciencias Básicas, Universidad de Pamplona, Pamplona, Colombia. [4]Unidad de Microbiología Aplicada, Instituto Canario de Investigaciones Agrarias, Santa Cruz de Tenerife, Spain.

Abstract: Fungi have been used to elaborate fermented foods by many cultures throughout the world since ancient times. Many filamentous fungi are currently used in food processing both in eastern and western countries. As food processors, filamentous fungi participate in different aspects during the elaboration of the product, contributing to its ripening, avoiding the growth of undesirable molds and bacteria, and providing the product with characteristic flavors and colors. *Penicillium* is the fungal genus most frequently found on foods, and some species are used in the food industry for elaboration of different products. This review will focus on three *Penicillium* species widely used in the western world as starter cultures/food processors. *P. nalgiovense* is frequently isolated from meat products, mainly dry fermented sausages. In recent years, this species has acquired biotechnological importance in the meat industry, due to its widespread use as starter culture. After inoculation with a suspension of asexual spores, *P. nalgiovense* develops on the product surface a typical homogeneous whitish thin layer of mycelium during the curing/ripening process. This mycelium contributes significantly to the product flavor, regulates moisture loss, and prevents the development of potentially mycotoxigenic fungal species. *P. camemberti* is the ripening agent of Camembert and Brie cheeses. It produces several enzymes, including exo- and intracellular proteases, some of which involved in casein hydrolysis during cheese ripening and thus contributing to the development of flavor and texture, and an alkalyne lipase able to hydrolyze mono- and diacylglycerols, which produces free fatty acids that are important for the organoleptic properties of these cheese varieties. *P. roqueforti* is one of the most frequently isolated fungus on cheese due to particular physiological features that allow it to colonize the inner part of the product, and plays a crucial role in the ripening of the so-called blue cheeses. Its strong proteolytic and lipolytic activities are responsible for the typical organoleptic features of these cheeses. In addition, during the ripening process *P. roqueforti* produces andrastins (A-D), compounds with antitumoral properties which inhibit the protein farnesyltransferase enzyme.

INTRODUCTION

Fungi have been used as food since ancient times by many cultures. In addition, some fungi were used for their special properties as narcotics or for medical uses, sometimes with a religious significance [1-3]. The gathering of fungal fruit bodies is probably the oldest use of fungi as food.

Mushrooms were prized by the ancient Egyptians and Romans. Records of the consumption of fungi in China date back to between 26 BC and 220 AD, but there is evidence of use as early as 6000 years ago [4].

The *sclerotium* of *Polyporus mylittae*, which became known as "native bread" or "blackfellows bread" and may weigh several kilograms, was consumed by Australian aborigines [5]. The *sclerotium* of *Poria cocos*, known as stone fungus or 'tuckahoe', is believed to have been roasted as a 'bread' and consumed by North American natives and peoples in Eastern Asia [6]. And *Ustilago maydis* has been reported to be eaten by native people in México [7].

Although several hundreds of species of edible mushrooms exist in the wild, less than 20 species are currently used extensively as food, and only 8-10 species are regularly cultivated to any significant extent [8]. *Agaricus bisporus* is

*Addres correspondence to Francisco Fierro: Universidad Autonóma Metropolitana-Iztapalapa, Edificio S-164, Avda. San Rafael Atlixco 186, Colonia Vicentina. Delegación Iztapalapa. 09340 México D.F. Tfno: +52 55 58046453. Fax: +52 55 58044712. fierrof@xanum.uam.mx

the most commonly edible mushroom cultivated in the world. *Pleorotus* spp. is another major crop around the world, especially the species *Pleorotus ostreatus*. *Lentinula edodes* (grown on logs) and *Volvariella volvacea* (on rice straw) are major mushroom crops in Japan and South-East Asia [9], and *L. edodes* has become popular also in Western countries and is now cultivated in the USA and European countries. Whereas in Western countries fungi are generally just a complementary product in diets, in some African countries they may account for up to 10% of daily protein intake.

Mycoprotein is a more recent development that uses fungi grown in fermenter vessels to produce "single-cell protein" for human consumption, mainly as additive for different processed or oven-ready meals. Although growth rates and yields of filamentous fungi are lower than most bacteria and many yeasts, harvesting of mycelium from growth media is easier and levels of nucleic acids are lower. A strain of *Fusarium graminearum* isolated from a soil sample was the first filamentous fungi to be used for mycoprotein production. Mycoprotein has some advantages over protein from animal or milk origin; it has a high-protein but low-fat content, this fat is rich in polyunsaturated fatty acids and is cholesterol-free (fungi have different sterols from those of animals), and has also high levels of dietary fibre and a low degree of allergenicity [10]. Mycoprotein can be dried and rehydrated, canned or frozen and it can be incorporated into a range of foods including soups and fortified drinks, biscuits, chicken-, ham- and beef-flavored protein, game pie and some fish products.

Fungi in the Elaboration of Food Products

Fungi have also been used to elaborate fermented foods by many cultures throughout the world. Baking, brewing and wine fermentation make use of the yeast *Saccharomyces cerevisiae*, but many other processes utilize filamentous fungi in different ways and at different steps of the processing to elaborate food products. Campbell-Platt and Cook [8] compile up to 30 types of mold-fermented foods. These mold-fermented foods can be classified according to the type of food and the participating microorganisms, and can be grouped as follows: i) cheeses ripened with molds, ii) sausages and *salamis* fermented with lactobacilli and ripened with molds, iii) salted smoke- or dry-cured ham ripened with molds, iv) proteinaceous plant foods initially fermented with molds followed by yeast and lactobacilli in the presence of high salt concentrations, v) proteinaceous plant foods made with *Rhizopus* or *Aspergillus* with or without salt.

The majority of molds involved in these fermentations belong to the genera *Aspergillus*, *Penicillium*, *Rhizopus*, *Amylomyces* and *Mucor*. Species of *Neurospora*, *Monascus* and *Actinomucor* are also involved in some fermentations. Species of *Aspergillus* and *Penicillium* are typical of products from temperate areas (cheese, meat, soy sauce), whereas *Rhizopus*, *Amylomyces* and *Mucor* are typical of products in predominantly tropical regions (tempe, tape). The diversity of food products produced by fermentations involving molds is considerable, ranging from mold-ripened cheese and meat, savoury staple foods such as tempe, flavoring products like soy sauce, and sweet foods including tape and brem cake [8].

Fermented foods such as mold-ripened meat and cheese are familiar in Europe and North America, but the majority of fermented foods involving filamentous fungi are produced in East and South-East Asia. In Africa, fermented foods are produced almost exclusively using bacteria and yeasts, which may be related to climatic and cultural differences, or to the limited use of rice and soybean in Africa.

An important difference exists in the perception and acceptance of molds in relation to food between Western and Eastern countries. In the West, the growth of molds on food is habitually regarded as spoilage, and often connected with the idea of "toxic". Only a few number of products ripened with molds are accepted and commercialized. In contrast, in China and South-East Asia, a great number of mold-fermented foods are known and regularly consumed [11-13]. In Western countries, almost exclusively species of the genus *Penicillium* are used in ripening processes of products such as Camembert and blue cheeses, dry-fermented sausages and smoke/dry-cured hams, all of them from animal origin, whereas different species of *Aspergillus*, *Mucor*, *Rhizopus* and *Neurospora* are used in Asia in the elaboration of foods of plant origin such as soy beans, rice and wheat [14].

Among the most popular fermented foods of plant origin we can mention soy sauce, miso, tempe, oncom and sufu. Soy sauce can be prepared with no or only small amounts of wheat (Chinese type) or with a 1:1 ratio of soy bean and wheat (Japanese type). The elaboration of soy sauce comprises two stages: the *koji* stage and the *moromi* stage

[15]. The *koji* stage is an aerobic solid substrate fermentation; soaked and cooked yellow soy beans mixed with roasted wheat are fermented with a mixture of molds termed *koji*, which is composed of yellow-green Aspergilli. During the *koji* stage, enzymes as amylases, proteases, cellulases, pectinases, invertase and others are released from the fungi and hydrolyze the soy bean and wheat substrates. The *moromi* stage consists of a submerged fermentation in which the hydrolytic products from the *koji* stage are substrates for the lactic acid bacterium *Teragenococcus halophilus*, and subsequently for the yeasts *Zygosaccharomyces rouxii*, *Candida versatilis* and *Candida etchellsii*, which form most of the flavor components of the sauce [15].

Other Oriental fermented foods of plant origin also make use of both fungi and *bacteria* in two-step fermentations, as in the case of *miso*, a Japanese fermented paste from soy bean, rice and barley or rye; here *Aspergillus oryzae* is responsible for the *koji* stage, in which fermentable products are formed to be used later as substrates by yeasts and lactobacilli. In other cases, the *koji* process itself serves as a single-step solid substrate fermentation for the production of foods like *tempe* (which uses *Rhizopus* spp.), *oncom* (*Rhizopus oligosporus* or *Neurospora sitophila*, depending on the variety) and *sufu* (*Actinomucor elegans* and others) [15].

In comparison to East and South-East Asia, the use of filamentous fungi for elaboration of food products in Western countries is restricted to a few species of the genus *Penicillium*, although species of *Aspergillus*, *Eurotium* and some other molds may also eventually appear during the maturation process of mold-ripened cheeses and meats.

Penicillium spp. as Ripening Agents for Cheese and Meat Products; Benefits and Potential Risks

Penicillium species constitute the main part of the mycobiota of fermented mold-ripened sausages (like *salami*), raw dry- or smoke-cured meat products (like ham), and several types of cheeses, such as Camembert, Brie and the so-called blue cheeses, contributing in different ways to the ripening of the product.

Penicillium nalgiovense is the most frequently isolated species from mold-ripened meat products. Other species, such as *P. chrysogenum*, *P. commune*, *P. solitum*, *P. nordicum* by *P. expansum*, and a few *Aspergillus* species are also habitual in the mycobiota of these products. Traditionally, in raw dry- or smoke-cured products, inoculation occurs naturally from the "house mycoflora" of the processing plant, whereas in fermented mold-ripened sausages, the product is inoculated with starter cultures to avoid growth of undesirable molds and ensure full colonization by the suitable strain. Mintzlaff and Leistner [16] isolated in 1972 the first *P. nalgiovense* strain that was used as starter for meat products. Strains from other species have also been used; Hwang *et al.* [17, 18] selected strains of a white variant of *P. chrysogenum* and strains of *P. camemberti* as starter cultures for dried sausages.

All blue-cheeses are made with *P. roqueforti*, which is inoculated and grows inside the product. *P. camemberti* is used in mold-ripened soft cheeses like Camembert and Brie, it is inoculated together with a lactic started culture or alternatively sprayed on the product, and grows on the surface of the cheese piece. Other *Penicillium* species may eventually colonize these cheeses and can also be isolated.

A starter culture consists of a microorganism or a combination of different microorganisms which contribute to the final organoleptic properties of a processed food. These microorganisms are able to transform the basic components of food (carbohydrates, lipids, proteins and others), resulting in positive changes in texture, organoleptic properties and, in many cases, long-time conservation. Additionally starter cultures may produce beneficial effects on the consumer health (probiotic foods). Thus, starter cultures have been subject of intense research in the last years [19].

Among the main benefits provided by the use of *Penicillium* spp. as starters on cheese and meat products we can mention: i) their contribution to the ripening of the product through enzymatic activities acting mainly on lipidic and proteinic substrates of the product, ii) their contribution to the organoleptic properties of the product as a consequence of these activities, iii) the antagonist action against other undesirable microorganisms, iv) protecting the surface of the product from dryness, light radiation and oxygen, thus avoiding that it becomes rancid. All these aspects will be discussed in more detail later for the individual species.

However, in some cases the presence of molds on food products should be avoided for several reasons: 1) risk of mycotoxin production [20], 2) rejection of the consumer, 3) development of unpleasant odours, appearance and changes in taste and nutritional value [21, 22]. In the meat industry, a number of methods have been developed to control the presence of molds, including the use of different antifungal compounds allowed in the food industry (e.g.

potassium sorbate) and mechanical removal of the mycelium prior to marketing (e.g. brushing or washing by application of water pressure). Subsequently, the products are protected by preservation methods (controlled or protected atmosphere, vacuum packaging or cold storage) throughout the distribution chain.

Penicillium species are known to produce a high abundance of secondary metabolites (SM). *P. chrysogenum, P. nalgiovense, P. griseofulvum, P. dipodomys, and P. flavigenum* produce penicillin [23], an antibiotic widely used to treat bacterial infections. But many other SM produced by *Penicillium* spp. fall within the category of mycotoxins, such as ocratoxin A, citrinin, verrucosidin, patulin, penicillic acid, cyclopiazonic acid or roquefortine C, which have different degrees of toxicity on target organs (nervous system, kidney) or are carcinogenic. In a study by Leistner and Eckardt [24] it was found that 80% of 1,481 *Penicillium* isolates from mold-ripened meat products were potential mycotoxin producers. For a detailed description of *Penicillium* mycotoxigenic species see [25] and [26].

The main concern regarding the use of fungi as ripening agents is the possible production of mycotoxins on the products. However, this risk is actually very low and is avoided by different strategies. The level of toxicity of mycotoxins on food is one million times lower than that of the botulinic toxin, and one hundred times lower than many toxins from algae. Consequently, the number of confirmed cases of intoxication by mycotoxins in humans is extremely low as compared, for instance, with intoxication by *Salmonella* or other infectious diseases. Pestka [27] concluded that even though molds isolated from fermented meat products have the potential to produce mycotoxins under appropriate conditions in the laboratory, this ability is probably very limited when growing on the meat products, and scant evidence exists that the market-ready products contain dangerous concentrations of mycotoxins. The presence of antibiotics like penicillin, produced by *P. nalgiovense* also on fermented sausages during the first stages of ripening [28], should also be avoided and will be discussed below.

There are different ways to help to combat the presence of mycotoxins and antibiotics on food. First, the use of a non-mycotoxigenic starter culture that can effectively compete with other mycotoxigenic fungi avoiding their growth. *P. nalgiovense* strains habitually used as starters do not produce mycotoxins [26], and several authors [29, 30, 31] have demonstrated the ability of *P. nalgiovense* to control the development of other mycotoxigenic fungi when applied as starter on cured-meat products. Strains of *P. camemberti* which does not produce cyclopiazonic acid have been isolated [32]; this is the only mycotoxin reported to be produced by this fungus. In the case of *P. roqueforti*, the production of several mycotoxins has been described [33]. Today, starter cultures are available which do not have the ability to produce most of these mycotoxins, and the presence of roquefortine C and isofumigaclavines in the amounts usually found on cheeses does not seem to represent a risk for human health in the least [34]. Production of mycotoxins can additionally be controlled with the use of genetic engineering techniques [31, 35, 36], for instance by disruption of genes specific for particular secondary metabolite pathways [31]. Genetic engineering may also be applied to improve other characteristics of fungal strains that will be used as starters, as the ability to antagonize bacterial growth [37-39].

Recently, a global regulator of fungal secondary metabolism, the methyltransferase LaeA, has been characterized in *A. nidulans* [40, 41]. Genomic analysis of *laeA*-deleted mutants has allowed identification of new putative secondary metabolite clusters [42, 43]. LaeA seems to be conserved in all secondary metabolite producing fungi, and has also been described in *P. chrysogenum* [44]. LaeA should be considered as a new target to prevent mycotoxin production. Inactivation of the *laeA* gene can potentially suppress the production of most of the SM in a fungus, however the resulting phenotype might not be suitable in all cases for starter cultures, as other characteristics such as asexual development may also be affected [44, 45]. In addition, deletion or attenuation of *laeA* does not cause a total suppression of the capacity to produce SM in *A. fumigatus* [43] and *P. chrysogenum* [44]; in the latter, roquefortine C production was not affected by RNAi-mediated attenuation of *laeA*, whereas penicillin yield showed a drastic decrease. Genome sequencing of *Penicillium* strains used as starters will provide information on their full mycotoxigenic potential, and will allow to increase our knowledge on the genes responsible for lipolytic, proteolytic and other enzymatic activities involved in the ripening of the products, thus allowing us to design strategies to improve their capabilities as starter cultures.

Excellent reviews have been published dealing with the microbiological aspects of different types of food fermentations [46, 47, 48] or focused specifically on mold-fermented foods [8, 15]. Next we will focus on the three *Penicillium* species most widely used in the western world in the elaboration of mold-ripened foods: *P. nalgiovense,*

P. camemberti and *P. roqueforti*. Microbiological, physiological and biochemical aspects will be covered, describing the benefits provided by the fungi on the products elaborated with them, the enzymes involved in ripening and contributing to the generation of organoleptic properties, and how in some cases they effectively compete with undesirable microorganisms preventing their growth. Other potential biotechnological applications of some of these species will also be discussed.

PENICILLIUM NALGIOVENSE

Taxonomy, Characterization and Production of Secondary Metabolites.

Penicillium nalgiovense was described by Laxa in 1932 from an isolate obtained from Ellischauer cheese. This cheese is similar to Camembert and is typical of the Nalžovy region, southern Bohemia (hence the specific epithet). The production of this cheese dates back to 1880. During the maturation process, a velvety greyish green mycelium covers its surface. *P. nalgiovense* can produce reddish-orange areas that, eventually, appear on the cheese surface. According to Laxa, *P. nalgiovense* is the predominant species during the maturation process. Raper and Thom [49] included *P. nalgiovense* in the Asymmetrica-divaricata series of *P. canescens*. Pitt [50] described the synonymy between *P. jensenii* Zaleski and *P. nalgiovense*, and the latter name was accepted and included in the genus *Penicillium* subgenus *Penicillium*.

Different isolates of *P. nalgiovense* develop colonies with visible morphological differences on solid culture media. Therefore, they were divided into six biotypes according to the diameter of the colonies, sporulation rate and color [51]. Nevertheless, these biotypes could not be differentiated neither by multivariate analysis of the production of secondary metabolites (nalgiovensin, nalgiolaxin and chrysogine) [52], molecular analysis by RAPD (Random Amplified Polymorphic DNA) [53, 54, Laich, unpublished data] nor ribosomal DNA and β-tubulin gene sequencing [54]. However, minor differences were detected in different isolates by fingerprinting analysis of *Eco*RI-digested total DNA with a (GTG)$_5$ probe [28].

P. nalgiovense produces a considerable amount of red pigment when the mycelium develops in culture media with high sodium or protein concentrations. These pigments appear on cheese and in the inner side of the skin of fermented dry sausages colonized by this fungus. Several authors managed to extract different pigments: nalgiovensin (orange crystals) and nalgiolaxin (yellow crystal) [55-57]. Subsequently, a large number of bioactive extrolites (secondary metabolites), such us citreoisocoumarins, were detected (see a detailed analysis in [26]).

Penicillin is another important secondary metabolite described in this species [58, 59]. *P. nalgiovense* carries the three structural genes involved in biosynthesis of penicillin, which are homologous to the respective *P. chrysogenum* genes [58]. The organization of the penicillin biosynthetic genes (*pcbAB*, *pcbC*, and *penDE*) is the same as in *P. chrysogenum* and *Aspergillus nidulans* [28]; the *pcbAB* gene is located in opposite orientation to the *pcbC* and *penDE* genes in all three species. Penicillin production in liquid complex medium ranged from 6 to 38 µg.ml^{-1} depending on the biotype and the particular isolate analyzed [28], as well as the sodium concentration of the medium [31]. Some differences were detected in the expression and regulation of the penicillin genes between *P. nalgiovense* and *P. chrysogenum* [60].

The karyotype of *P. nalgiovense* and *P. chrysogenum* was determined by pulsed-field gel electrophoresis, showing that both contained four chromosomes. The estimated genome size of *P. nalgiovense* was 26.5 Mb, whereas in *P. chrysogenum* it was 34.1 Mb. The penicillin biosynthetic genes were located on chromosome IV (4.1 Mb) in *P. nalgiovense* [61], whereas in *P. chrysogenum* they were located on chromosome I (10.4 Mb) [62].

The *naf* gene encoding an antifungal protein has also been identified in *P. nalgiovense*. This gene is homologous to the *paf* gene of *P. chrysogenum*. When *P. nalgiovense* was grown in Petri dishes containing other fungal species, a weak antifungal activity was detected [63].

Concerning mycotoxins, this species apparently does not produce this type of compounds [26, 64]. However, it is interesting to note that in some strains, a weak PCR product was detected using primers specific for two genes of the ochratoxin A (OTA) biosynthetic pathway: a polyketide synthase "otapksPN" and a non ribosomal peptide gene syntethase "otanpsPN". Nevertheless, none of these strains produced detectable amounts of OTA. A more detailed analysis revealed that *P. nalgiovense* carries non-transcribed sequences similar to the OTA biosynthetic genes of *P. nordicum* [65].

The Mycobiota of Meat Products

The development of a surface mycobiota on meat products is mainly due to the existence of an environmental "house flora" in the manufacturing plants. This contamination occurs at different stages of manufacturing, and fungal development occurs only under favourable conditions. These conditions vary for each fungal species, and their adaptability determine the colonization capacity. The curing time of meat products in the different ripening stages is correlated with the level of humidity and water activity, which directly affect the composition of the mycobiota. Mold growth on the surface of dry cured meat products is the result of their tolerance to low pH and high salt concentration [66].

Fungal development on meat products is evident by the surface layer of mycelium. This layer avoids moisture loss and protects the product from light and oxygen, preventing it from becoming rancid. The microclimate created on the surface prevents the formation of an oily and dry layer, and may be beneficial for the development of characteristic aroma and flavor through the production of proteases and lipases [67]. Several authors describe the important role of fungi during the maturation process through the proteolytic, lipolytic and anti-oxidation processes [67-82]. Also, certain desirable effects have been detected in the formation of flavor, aroma or texture of the finished product [74, 77, 83, 84] and even fungi can act as a "biocontrol" agent against other unwanted organisms [29, 30].

The development of filamentous fungi on the surface of meat products was initially studied by Carbone [85]. Later, several authors described the changes in the composition of the mycobiota associated with different cured meat products, such as fermented dry sausages, hams in Spain, bündnerfleisch in Switzerland, bauernspeck or speck in Austria and Italy, Bresaola in Italy and biltong in South Africa, among others [16, 21, 51, 69, 71, 82, 86-114]. Usually, *Penicillium* species constitute the predominant mycobiota in products with a short ripening period (e.g. *salami*), whereas in products with a longer ripening period (e.g. ham) *Penicillium* species dominate the beginning of the process and then are displaced by xerophilic species of *Penicillium, Aspergillus* and *Eurotium*. The dominance of certain species increases the degree of environmental contamination and generates a constant natural inoculum in the ecological niche where they thrive.

P. nalgiovense has been reported as associated with different dry cured meat products [21, 22, 71, 82, 108, 109, 113, 114, 115] and it can also be found in the air of the manufacturing plants [113]. Interestingly, in Norwegian dry cured meat products smoking has a selective effect on different species of *Penicillium*, causing an increase in the isolation frequency of *P. nalgiovense* and a reduction in the occurrence of *P. solitum* and *P. commune*. Smoking seems not to have a preventive effect on the growth of *P. nalgiovense,* which was the dominant species isolated from both smoked and non-smoked products [116]. Enzymes were reported to contribute to lipolytic and proteolytic activities that generate flavor precursors and improve the texture [52, 73, 76-80].

P. nalgiovense in the Food Industry

The first references in the literature on the use of yeast strains for the production of dry sausages are those of Cesari [117] and Cesari and Guillermond [118]. However, the selection of filamentous fungi strains to be used as starter cultures in meat products was not implemented until 1972, when Mintzlaff and Leistner [16] selected a strain of *P. nalgiovense*. Currently, different strains of this species are marketed by various companies and used extensively in different meat products around the world.

The application of *P. nalgiovense* conidia on the surface of meat products enables the development of a typical mycelium of white or green color (depending on the biotype) that contributes to the formation of the flavor and prevents the development of undesirable species. The best starter culture should be defined by each manufacturing plant. The inoculation of the surface of meat products should be done with species of fungi that have been selected for this purpose. It is important to note that there is an appreciable difference in the conidiogenesis rate and response to different substrates among different isolates of *P. nalgiovense* [119]. Temperature, water activity (a_w) and NaCl concentration are the most influential factors on the growth rate of *P. nalgiovense*. The increase of NaCl to a concentration of 4% produces a significant stimulus to the growth rate and nalgiovensin synthesis of *P. nalgiovense*

[119]. Therefore, the selection of the most appropriate strains is very important. This will ensure a proper colonization of the product, and will avoid the development of potentially mycotoxigenic species or species that confer undesirable flavor [29, 30, 120, 121].

Inoculation is usually done by immersing the product in a suspension of about 10^6 conidia.ml^{-1} of water. In other cases, it is made by spraying a conidial suspension on the surface. The correct application of the starter culture is the key to a good colonization of the product and to obtain an efficient biocontrol of the undesirable fungal species. In Fig. **1** it is shown the appearance of different meat products inoculated with a selected strain of *P. nalgiovense* and a control without inoculation.

P. nalgiovense is use as starter mainly on dry sausages, but it can be applied to other meat products. Laich *et al.* [31] demonstrated the ability of a native *P. nalgiovense* strain to colonize *cecina* (a salted smoke-cured beef meat product from the region of León, Spain). In this case, conidia (a suspension containing 5.6×10^6 conidia.ml^{-1}) were sprayed at the beginning of the ripening process (just after salting the meat pieces) on the surface of pieces of cecina. The development of the mycelium layer exerted a very good control on the growth of other fungal species. In contrast, the non-inoculated pieces were colonized by the diverse typical mycobiota growing on cecina (Fig. **1**).

A problem associated with the use of *P. nalgiovense* as starter is its ability to produce penicillin [58, 59], which has been shown to occur also on the surface of products like *fuet*, a fermented sausage from Spain [28]. Significant amounts of penicillin were found *in situ* in the casing and the outer layer of salami meat during early stages of the curing process, coinciding with fungal colonization, but no penicillin was detected in the cured salami. The presence of antibiotics like penicillin in food is not convenient, due to the fact that it may produce allergic reactions [122-124] and lead to the development of resistance in resident bacteria in humans, which may eventually transfer that genetic information to pathogenic bacteria [125]. Laich *et al.* [31] obtained *P. nalgiovense* strains impaired in penicillin production by disruption of the *pcbAB* gene (the first gene of the penicillin biosynthetic pathway).

Figure 1: Development of surface mycobiota. 1; cured meat product (*cecina*) after 42 days, post-salting stage. 2; different types of fermented dry sausages after 10 days of ripening. A; inoculated with *P. nalgiovense*, B; non inoculated and colonized with different fungal species of the "house flora".

When applied as starter on smoke-cured beef meat (*cecina*), the *pcbAB*-disrupted strain showed no differences with respect to the parental penicillin-producing strain in its ability to colonize the meat pieces and to control their normal mycobiota. Both strains also carried out a similar control on the presence of bacteria in cecina. A similar proportion of penicillin-sensitive and penicillin-resistant bacteria were isolated from pieces inoculated with the penicillin-producing or the non-producing *P. nalgiovense* strains. These results suggested that the decrease of bacterial population on the surface of cecina was caused by the higher competition for nutrients due to the development of *P. nalgiovense* mycelium, and not for the production of penicillin by this fungus.

Genetic transformation of *P. nalgiovense* has been achieved by several researchers. With this technology, some heterologous genes have been introduced in *P. nalgiovense* strains, as the lysostaphin gene of *Staphylococcus staphylolyticus* [126] and the glucose oxidase gene from *A. niger* [38, 39, 127]. In both cases, expression of the heterologous genes provided the transformed strains with antagonistic activity against foodborne pathogenic bacteria. Fierro *et al.* [36] described a series of vectors, selectable markers and transformation methods that can be used for efficient transformation of *P. nalgiovense*, gene cloning and expression. These tools greatly facilitate the investigation at the molecular level with the aim of characterizing the biosynthetic pathways of interest in the food industry.

In most cases, products inoculated with *P. nalgiovense* are marketed with the layer of mycelium. The fuet is a typical case of this kind of products in Spain. In fuet, the marketing is done in two different ways, without packaging or packaged. The first is the traditional case, but the period during which the product can be commercialized is lower due to the loss of moisture. Therefore, with the aim of prolonging the shelf life of the products while maintaining quality, the use of different methods of conservation has been increased significantly in the last 10 years. Indeed, most dry sausages are packaged today in modified atmosphere (Modified Atmosphere Packaging, MAP) or vacuum. In the case of inoculated sausages marketed with the layer of *P. nalgiovense* mycelium on the surface, the use of these packaging systems is not possible. Exposure to an atmosphere of oxygen lower than 5% or with a CO_2 concentration above 15% leads to the death and degradation of the mycelium, although the viability of conidia is not affected [128]. In sausages wrapped in a modified atmosphere (70% N_2 and 30% CO_2), oxygen remaining values are less than 1%. So, considering the consequences of this concentration of oxygen to the mycelium of *P. nalgiovense*, sausages with surface mycelia should be packaged in microperforated bags. This package system allows oxygen exchange with the outside air, but at the same time creates a microenvironment with higher water content, increasing the shelf life of the product.

Finally, it is important to emphasize that the choice of strains to be used as starter cultures in food should be made from microorganisms naturally adapted to the conditions of ripening and developing of the product. These strains contribute to the organoleptic characteristics of the product, and should be antagonists of undesirable microorganisms.

PENICILLIUM CAMEMBERTI

One of the most used starters in the food industry is *Penicillium camemberti* (Fig. **2**), the ripening agent of Camembert and Brie cheeses, to which it contributes many of their organoleptic properties.

In addition to cheeses, *P. camemberti* has been tested recently on dry fermented sausages to improve their organoleptic properties during the ripening process, yielding interesting results [129].

Figure 2: *P. camemberti* growing on the surface of a Camembert-type cheese.

Here we summarize our current knowledge on the enzymatic and physiological properties of *P. camemberti* as food ripening agent, as well as its potential as biotechnological model for the production of other interesting products. It is important to remark that some data here described are taken from publications where *P. camemberti* was named *P. caseicolum*. However, since the end of the 70's, different taxonomic re-evaluations resulted in a majority consensus that both belong to the same species, differing only in spore color [130-132]. Thus, for the purpose of this article, we will use the name *P. camemberti* in the whole text.

Overview of the Contribution of *P. camemberti* to the Ripening of Soft Cheeses

The organoleptic properties of food are produced by an intricate and dynamical net of biochemical reactions involving hundred of compounds and intermediates with different properties. In the case of cheese, these biochemical reactions are part of the ripening process, in which several microorganisms may participate [133]. *P. camemberti* is one of the most important contributors to the organoleptic properties of some kind of cheeses. Several experimental data indicate that inoculation of *P. camemberti* produces changes in organoleptic properties of cheeses and other foods [134, 135].

At the beginning of the 20th century, biochemical studies of *P. camemberti* aiming to characterize the enzymatic reactions involved in the ripening of cheeses were initiated. The first studies were focused on the determination of intracellular activities of *P. camemberti*, such as protease, lipase, nuclease, amidase, and glycosyl hydrolases, among others [136]. Extracellular activities were not studied at that time since it was believed that most organisms did no secrete their enzymes to the extracellular media [136]. Today, it has become evident that extracellular enzymatic activities are responsible for most of the organoleptic properties produced by *P. camemberti*.

The contribution of these extracellular activities from *P. camemberti* to ripening is directly related to the environmental conditions that the fungus meet, which influence its growth, mainly on the surface of soft cheeses. At early stages of maturation of soft cheeses, residual lactose from milk is quickly metabolized to lactate by the bacterial population. Lactate reaches concentrations of about 1.5% at 24 hours, and produces a high acidification of the cheese dough. Lactate constitutes an abundant carbon source for organisms such as *P. camemberti* [133, 137, 138]. The catabolism of lactate by *P. camemberti* produces CO_2 and water, and causes a pH gradient across the cheese piece, with the concomitant migration of Ca^{2+} ions and lactate from the inner to the surface of the cheese, thus producing its softening [133]. In addition, this process causes an increase of the pH on the surface of cheese, which allows the activation of some inactive proteases. Once lactate is exhausted, *P. camemberti* is forced to use other secondary carbon sources, which correlates with an increasing importance of the lipolytic and proteolytic activities, the main extracellular enzymatic activities from *P. camemberti* involved in the ripening of cheeses.

Extracellular Lipolytic Activity from *P. camemberti*

Milk fat metabolism is necessary for the development of the flavor during the ripening of cheese, being essential for this process the enzymatic hydrolysis of triglycerides to fatty acids and glycerol, mono- or diglycerides [133]. In the case of Camembert cheeses, 95% of free fatty acids are originated from lipolysis of milk fat [131]. This degradation occurs mainly by the action of a variety of lipases from different origin acting on cheese. One of these lipases is produced by *P. camemberti*.

P. camemberti possesses a potent extracellular lipase activity. As an example, *P. camemberti* U-150, an industrial strain used for lipase production, was isolated from a soil sample by a screening looking for microorganisms with high lipase activity [139]. From culture filtrates of *P. camemberti* U-150, at least two to four lipase active fractions were initially described [139, 140]. However, after more detailed experiments, it was clear that these different "forms" were the same enzyme with different composition of carbohydrates attached to the protein, which alters its size and migration properties. Thus, so far, only one extracellular lipase has been isolated from *P. camemberti*.

The protein sequence of *P. camemberti* lipase (deduced from the nucleotide sequence of its encoding gene) corresponds to a 305 aa protein, with a putative signal peptide of 26 aa, thus giving a mature protein of 279 aa [139]. However, further direct peptide sequencing showed a slight difference in size, with a mature polypeptide chain of 276 aa [141]. From these data, a putative catalytic triad involving Ser[145], Asp[199] and His[259] was postulated [139, 141]. Further site-directed mutagenesis experiments confirmed the role of Ser[145] and His[259], and additionally Ser[83], but although Asp[199] is also likely to be involved, its putative role in catalysis has not been totally proved [142].

At the biochemical level, the main characteristic of *P. camemberti* lipase is its substrate specificity. This lipase has no activity on triacylglycerols, and is strictly a mono- and diacylglycerol lipase [139], which is a rare case among fungal lipases. In fact, this is the only microbial lipase with these characteristics described so far. However, and given that the sequence and structural characteristics of the *P. camemberti* lipase resembles fungal triacylglycerol lipases [139], this enzyme is commonly grouped with triacylglycerol lipases, which are mainly found in the E.C. 3.1.1.3 group and are, strictly, triacylglycerol acylhydrolases.

P. camemberti lipase has been modeled and crystallized [140, 143], concluding that this enzyme belongs to the α/β hydrolase superfamily, with high structural similarity to triacylglycerol lipases from filamentous fungi. Isobe *et al.* [140] proposed that differences in the specificity of substrate are due to differences in its pocket, specifically in the region close to reactive His^{259}, which seems to be determinant for the substrate preference for monoacylglycerols and diacylglycerols.

Although other microbial lipases present in cheese could be able to hydrolyze milk triglycerides to provide *P. camemberti* lipase with substrates in the form of mono- and diglycerides, the presence of other extracellular lipases with activity on tryacylglycerols cannot be ruled out in this fungus. This kind of lipases has been described in other fungi with very close phylogenetic relationship to *P. camemberti* [144, 145].

Effect of Fatty Acid Metabolism of *P. camemberti* on Ripening

Upon action of lipases, free fatty acids are released, which are taken up by *P. camemberti* and metabolized, thus causing a noticeable contribution to organoleptic properties of cheese.

From lypolysis of milk fat, the main free fatty acids released are long chain fatty acids, mainly 16:0 and 18:1, with values around 1,500 mg/100 g of cheese [131]. These fatty acids can be metabolized by different pathways to produce several products that contribute to organoleptic properties: lactones, aldehydes (and consequently acids and alcohols) and, by the β-oxidation pathway, a set of oxidative volatile derivatives, generically named methyl ketones, which can be reduced subsequently to secondary alcohols by dehydrogenases [133, 134]. Both methyl ketones and its derivative secondary alcohols have been described as main contributors to organoleptic properties in cheeses.

P. camemberti readily reduces milk fatty acids, producing around 60% methyl ketones, mainly 2-nonanone and 2-undecanone [146]. The quantity of methyl ketones in cheeses inoculated with *P. camemberti* is around 25-60 µmol per 100 g of extracted fat from the Camembert cheese [131], but there are differences in the quantity and the specific type of methyl ketone produced, which are due to differences in strains and the fatty acid used as substrate. Experimentally, if pure caprylic and lauric acids are used instead cheese, mycelia from different *P. camemberti* strains produce several times more methyl ketones than those found in Camembert cheese [131], which is expected because of the use of pure compounds. Still more interesting, using these pure compounds, *P. camemberti* produces more heptanone than undecanone. In addition, and depending on the strain, amounts of produced heptanone range from 1.1 to 44.4 µmol per gram of pure fatty acids. Similarly marked differences can be observed for undecanone production [131]. On the other hand, Karahadian *et al.* [134] demonstrated that these differences also include the appearance of rare compounds. These authors demonstrated that some strains of *P. camemberti*, but not all the assayed strains, were able to produce the unusual compound 8-nonen-2-one, which has been hypothesized to be derived from a precursor produced by linoleic or linolenic acid metabolism. In addition to methyl ketones, volatile secondary alcohols have been also mentioned as important for organoleptic properties produced by *P. camemberti* in cheeses. Most of these volatile secondary alcohols are derived from the reduction of methyl ketones, and as happens with these compounds, their presence is highly variable, depending on the chemical precursors and strains used. For example, Jollivet and Belin [147] analyzed ten *P. camemberti* strains, detecting important differences in the production of volatile secondary alcohols and their respective methyl ketone precursors.

Production of secondary alcohols seems to be correlated to specific intracellular enzymatic activities of *P. camemberti*. For example, for the production of the volatile secondary alcohol 1-octen-3-ol from linoleic acid metabolism, enzymatic reactions involving activities such as lypoxigenase and hydroxyperoxide lyase are necessary [148, 149]. Actually, hydroperoxidase lyase activity is enhanced by linoleic acid [150]. Other factors, such as the composition of the culture media, also influence these enzymatic activities. Using a synthetic medium supplemented with different fatty acids and oils, Husson *et al.* [149] determined that the presence of soybean oil enhanced 8 times the production of 1-octen-3-ol with respect to the same medium containing only pure linoleic acid. On the other hand, these authors observed that when dairy medium was used instead of the synthetic medium, growth of *P. camemberti* and production of secondary alcohol were higher. However, these observations were not correlated with the enzymatic production of the same alcohol using enzymatic extracts obtained from the same cultures. Thus, enzymatic pathways for production of secondary alcohols seem to be a complex system that deserves a more detailed study.

Jollivet and Belin [147] proposed the grouping of some *P. camemberti* strains in two groups based on their aromatic characteristics: 1) those with fruity and spicy odor produced by a mixture of methyl ketones and their derived secondary alcohols, and 2) those with fruity and alcoholic odor, produced mainly by the most common secondary alcohols 3-methylbutanol and 2-methylpropanol. Thus, the marked differences in methyl ketones and secondary alcohols patterns in *P. camemberti* would make possible the selection of specific strains to produce cheeses with specific organoleptic properties.

Extracellular Proteolytic Activity from *P. camemberti*

Complex proteolysis events play a pivotal role in the development of organoleptic properties in cheeses. During ripening, degradation of casein is mostly produced by chymosin added as coagulant, and to a lesser extent by plasmin from milk. These proteases produce a huge quantity of peptides, some of which are degraded by these same enzymes, whereas others are degraded by proteases from the starter [133]. Similarly to *P. roqueforti* (see below), *P. camemberti* possesses a complete set of extracellular proteolytic enzymes:

1) Aspartyl protease: this is the most studied protease from *P. camemberti*. This enzyme is similar to other fungal acid proteases, with a molecular weight for the mature polypeptide of 33.5 kDa. It is active in a pH range from 2.5 to 6.5, with optimal activity at 3.4, and is inhibited by reagents specific to aspartic proteases [151].

2) Metalloprotease: One metalloprotease has been characterized in *P. camemberti* [152]. It was initially named neutral protease, and has an optimal pH between 5-7.5, depending on the incubation conditions and substrate used. The metalloprotease contains one atom of zinc per molecule, and its activity falls to 4% when is incubated with 10 mM EDTA [152, 153]. This enzyme clearly differs from the most studied bacterial metalloproteases, which show a smaller size, about 20 kDa, and a weaker specificity for casein [152].

3) Aminopeptidases: Two kind of aminopeptidase activities have been described in *P. camemberti* [131]. One of them, named "intracellular", is optimal at pH 8. However, it is not clear if this enzyme belongs to an unprocessed form of an extracellular aminopeptidase. Regarding extracellular aminopeptidases, a prolyl peptidase has been purified and characterized from culture filtrates of *P. camemberti* [132]. Different to other prolyl peptidases, this enzyme has a big size, with a molecular mass estimated to be 270 kDa, although when ran in a denaturing SDS-PAGE it shows two proteins bands of 66 and 83 kDa. The enzyme has optimal activity at neutral pH, and seems to be a sulphydryl enzyme, because it is inhibited by *p*-chloromercuribenzoic acid. Substrate specificity assays showed that the enzyme hydrolyzed dipeptides and some oligopeptides with proline in the amino end, but was inactive on peptides with proline at the carboxyl end [132]. Another aminopeptidase activity, described in general terms as "three active bands around 35 kDa", has been described in *P. camemberti* [131]. These three bands have optimal activity at pH 6-8 and show slight differences in aminopeptidase specificity. Due to their different characteristics, it does not seem likely that a prolyl peptidase described by Fuke and Matsuoka [132] belongs to one of these three "band activities", which suggest the presence of several extracellular aminopeptidases in *P. camemberti*.

4) Carboxypeptidases: According to Cerning *et al.* [131], *P. camemberti* has two carboxypeptidases. One of them (extracellular) has been purified, showing a molecular mass of 120 kDa and optimal activity at acid pH around 3-3.5. The other carboxypeptidase activity, associated to the myceliar fraction, has been described as a neutral activity, with optimal pH around 6-6.5. However, the enzyme(s) producing the latter activity has not been purified [131].

Finally, it is important to note that none of the genes encoding these proteases has been sequenced in *P. camemberti*. In fact, from cheese-ripening fungi, only one protease-encoding gene has been sequenced and published, the aspartyl protease from *P. roqueforti* [66].

Casein Degradation by Proteolytic Activity from *P. camemberti*

The study of the specific role of proteases from *P. camemberti* in casein degradation is carried out with aseptic model curds free of microorganisms, in order to avoid protease contamination from other sources. These artificial curds mimic the cheese environment. In these models, Desmazeaud *et al.* [155] monitored the casein breakdown produced by the inoculation of *P. camemberti*, and observed that this fungus degrades β casein and the large fragment $\alpha_{S1}I$ (α_{S1} casein lacking residues 1-23 from N-terminal) more efficiently than caseins γ and χ, causing a more intense degradation as compared to lactic bacteria. Using the same model, Gripon *et al.* [156] followed the specific effect of a neutral protease (currently the metalloprotease) in casein degradation, obtaining an almost identical degradation pattern to the former described using entire *P. camemberti*. A more detailed study using purified casein fractions demonstrated that, from a qualitative point of view, this metalloprotease acted in a similar way on α_{S1} and β caseins [157].

The aspartyl protease from *P. camemberti* was very active on β casein, with the enzyme splitting at least three specific bonds [158]. Less effective seems to be the action of this protease on α_{S1} casein; yielding a product of high electrophoretic mobility similar to the α_{S1}-I fragment described above. This result reflects the resistance of this casein form to the action of this protease, illustrating the specificity of cheese proteolysis [158].

Effect of Proteolytic Metabolism of *P. camemberti* on Ripening

The proteolytic activities of *P. camemberti* also contribute to the production of taste and aroma components. Short peptides and amino acids produced from proteolysis may suffer several metabolic reactions such as transamination, decarboxylation, and oxidative deamination, the last producing ammonia (another important component of cheese flavor) and α-keto acids, which are precursors of some aldehydes such as 2-methylpropanal and 2-and 3-methylbutanal [129, 131]. Bruna *et al.* [129] described that *P. camemberti* possesses an intracellular L-amino oxidase activity, which produces changes in the pattern of amino acids, ammonia content, volatile compounds and organoleptic properties of food.

In addition, it is important to note that secondary alcohols are not only produced by fatty acid metabolism. These compounds can also be produced by the specific metabolism of some amino acids. For example, 2-methylpropanol and 3-methylbutanol are derivatives from the metabolism of valine and leucine in *P. camemberti*, and methanethiol is produced by methionine catabolism [147].

Genetic Improvement and Potential for Recombinant Protein Production

As we have described, *P. camemberti* possesses an enzymatic battery that allows its use as ripening agent. These positive characteristics can be improved using different approaches, such as genetic engineering. For example, using suitable genetic tools, additional copies of lipase or protease genes could be incorporated, thus enhancing these important activities for ripening. In addition to genetic manipulation aiming improvement of its qualities as starter, *P. camemberti* can be suitable for other biotechnological applications. Together with the cheese, the mycelium of *P. camemberti* has been eaten by people for centuries without apparent health damage; for this reason, this filamentous fungus is recognized as a GRAS (Generally Regarded as Safe) organism. This gives *P. camemberti* a great advantage to become a potential producer of heterologous recombinant proteins that can be used as food additives or for pharmaceutical purposes [159].

Both genetic improvement and heterologous protein production require suitable genetics tools, as efficient transformation systems. Although *P. camemberti* can be successfully transformed using dominant antibiotic markers, this kind of markers are usually not accepted for biotechnological applications related to food organisms, and metabolic auto-markers are preferred, thus avoiding foreign DNA and antibiotics.

Whereas some attempts have been done in *P. roqueforti* to use it as recombinant protein producer [160, 161], to our knowledge there is no reports on the use of *P. camemberti* for such purpose. Amano Inc. (Japan) has patented a plasmid containing the hygromycin resistance gene under the control of the promoter and terminator sequences from the *mldA* gene, which encodes the lipase from *P. camemberti* [162]. In theory, the simple change of the hygromycin gene by a gene encoding a protein of commercial interest, and the co-transformation of *P. camemberti* with this new recombinant construct and a helper plasmid containing an environmentally friendly selection marker, would be sufficient to produce a recombinant protein in this fungus.

Other Potential Biotechnological Applications of *P. camemberti*

Fungal lipases are important for several industrial applications [163]. The *P. camemberti* lipase is a promising enzyme to be used as biotechnological tool. This enzyme is able to change its hydrolytic activity by the inversed glyceride synthetic reaction, maintaining its unique substrate specificity [164]. Thus, the action of this inverted enzymatic activity can increase triglyceride content in some types of oils and fats through the synthesis of triglyceride from diglyceride in a micro-aqueous environment. The same effect can be produced using the "classical" hydrolysis reaction. In this case, this activity achieves the specific and selective decomposition of diglycerides into glycerol and fatty acids, thus increasing the percentage content of triglycerides. In addition, this enzyme can be used in the synthesis of specific emulsifiers, which can be produced in this way at a high purity without further purification. Details about the use of *P. camemberti* lipase for these industrial purposes can be found at http://www.amano-enzyme.co.jp/eng/productuse/oil.html

P. camemberti possesses some other properties that could have potential biotechnological applications. For instance, some strains of this fungus produce the lactone antibiotic brefeldin A [165], although details about the physiology, genetics, or application of this secondary metabolite have not been described yet. On the other hand, there are several reports describing the degradation of toxic compounds from paper industry wastewaters by *P. camemberti* [166, 167]. Therefore, the application of this fungus as bioremediation agent is an attractive possibility.

PENICILLIUM ROQUEFORTI

Penicillium roqueforti is a filamentous fungus widely distributed in nature, as shown by the fact that it has been isolated from sources as diverse as water, soils, air, crops, processed and stored foods, and the environment of food industrial factories among others [168-176]. It is the *Penicillium* species most frequently found in a diversity of dairy products, especially in ripened cheeses [169, 177-180]. This behavior is due to its ready physiological adaptation to the physical-chemical and nutritional microenvironment of the casein curd resulting from the milk coagulation, which constitutes the base for cheese elaboration [180-184].

P. roqueforti has been reported to produce some mycotoxins, as PR-toxin, eremofortin, roquefortine C, mycophenolic acid, patulin, penicillinic acid and isofumigaclavins [177]. However, the amount of produced mycotoxins was found to be very different in laboratory cultures [185] and when grown directly on cheese [186, 187]. Whereas in laboratory media all the mycotoxins mentioned above can be found, in blue cheese samples only roquefortine C is regularly detected [177, 186-188]. Finoli *et al.* [187] found that roquefortine production levels on laboratory media were 4- to 6-fold higher than those detected on blue cheese. Similarly, when *P. roqueforti* was grown in milk at 4 °C and 20 °C, simulating the conditions of the ripening process, roquefotin production was significantly lower as compared to results obtained in laboratory cultures [187]. Actually, the substrate, temperature and microaerofilic conditions in which the fungus grows during the ripening process are determinant factors for the low roquefotin production levels [189, 190]. These low roquefortine levels and the relatively low toxicity of the mycotoxin make blue cheese a safe product for consumers [187]. As a consequence of this, most *P. roqueforti* strains are considered GRAS, and are used as starters for blue cheese production [177].

Due to its physiological characteristics, *P. roqueforti* is the microorganism specifically used for the ripening of the so-called "blue cheese", which owe this name to the color of the cheese dough provided by the conidia of the fungus [182, 191-194]. Among the most known blue cheeses we can mention Roquefort, Stilton, Gorgonzola, Cabrales, Amablu, Dolcelate, Bergader, Cambozola, Montagnolo, Gamonedo, Valdeón, Monje Picón, Gammelöst, Danablu, Fourme d'Ambert, Beenleigh blue, Edelpilzkäse, Kuflu and the French "Bleu", [180, 182, 184, 191, 193, 195-200].

P. roqueforti and the Microbial Ecology of Curds Prior to the Ripening Process.

In every one of the many existing blue cheeses, *Penicillium roqueforti* plays a main role. Its preponderance depends on the specific conditions of the elaboration of each product: the use of milk from cow, sheep or a mixture of different origins, if milk is pasteurized, the lactic microbiota typical of each area, the acidity of the curd, the type of bacterial starter used (if any), and other technological aspects [183, 201, 202]. Due to the many factors that come into play, a great variety of blue cheeses exist, each with particular organoleptic features that make it different from the rest [182, 194].

Most of the procedures for elaboration of blue cheeses include the inoculation of the milk or of the curd with conidia from *P. roqueforti* [203]. However, some hand-crafted manufacturing processes do not make use of any inoculation step [200], so that in these cases the fungi growing on the cheese come from the conidia present naturally in the curd along with the microbiota typical of milk [184]. In this case the bacteria and yeasts present are of vital importance, as they are the first to act generating the physical-chemical and nutritional conditions suitable for the later germination and proliferation of *P. roqueforti* [184, 202, 204].

The main bacterial microorganisms associated to the process of ripening of blue cheese belong to the gerera *Lactococcus*, *Lactobacillus*, *Leuconostoc*, *Streptococcus*, *Staphylococcus* and *Enterococcus* [200, 203, 205, 206]. In the case of yeasts, the reported genera are *Debaromyces*, *Candida*, *Saccharomyces*, *Cryptococcus* and *Kluyveromyces* [207, 208]. The genera *Yarrowia*, *Trichospora* and *Zygosaccharomyces* have been isolated from Danablu cheese [202]. Initial microbiota has two essential effects for the later development of *P. roqueforti*: (*i*) they reduce the pH, mainly due to the conversion of lactose in lactic acid and other volatile acids, and (*ii*) heterofermentative species produce enough CO_2 as to generate microcavities and grooves that allow diffusion of oxygen to the inner part of the cheese dough and will be the niche for germination of fungal conidia and the later colonization of the cheese [184, 202, 203]. The fungal colonization process may also be facilitated by perforating the curd, which is usually done after 2-3 weeks in Gorgonzola and Roquefort cheeses and after 5-6 weeks in the case of Stilton cheese [184, 197].

The subsequent salting step reduces humidity and helps to harden and make firmer the cheese surface prior to the ripening process. This step is the turning point for the bacterial population of the cheese, as the viable number of cells decreases considerably [193, 197]. The degree to which the initial microbiota contributes to the changes occurring during maturation is not very well known, but it is possible that the release of intracellular enzymes, together with the curd enzymes, contribute to the partial hydrolysis of casein in the first phase of ripening, just before *P. roqueforti* starts the active colonization of the cheese [192, 201, 209]. In addition, some yeasts may contribute to the lipolytic action in these initial phase [207, 210].

The differential distribution in time and space of the different microbial species suggests that ecological reasons and specific cooperative actions determine the places where growth will take place in the cheese, which will have significant consequences for the final characteristics of the product [205].

Physiological Characteristics Giving *P. roqueforti* an Adaptative Advantage

As mentioned above, *P. roqueforti* possesses particular physiological characteristics with respect to other fungal species, characteristics that allow it to develop successfully in the conditions generated by the accompanying microbial communities [177, 183, 211]. It must be taken into account that the microenvironment present in blue cheeses is characterized by pH and saline concentration gradients between the surface and the inner part of the cheese, which equilibrate progressively along the ripening process [83, 193].

P. roqueforti is one of the few fungal species considered as microaerophilic, which allows it to grow with low O_2 and high CO_2 concentrations and thus develop in the inner part of the cheese dough [169, 190, 212, 213]. It is one of the *Penicillium* species with lower requirement of O_2, growing usually at pressure values as low as 75 mm Hg [169]. Moreover, certain levels of CO_2 slightly stimulate germination of its conidia [184, 214]. *P. roqueforti* also tolerates well the salt concentrations used in the elaboration of blue cheeses; it is inhibited only at salt concentrations above 3-4%, and even its germination and conidiation are stimulated by NaCl concentrations around 1% [183, 184, 212, 215]. All these features, together with the fact that *P. roqueforti* shows an increase of its lag phase at a water activity (a_w) below 0.92, endow it with a competitive advantage to germinate and grow in the curds. One of the consequences of the salting process is that a_w decreases down to 0.91-0.94 [177, 216]. The a_w values tolerated by *P. roqueforti* to germinate and grow have led some authors to classify it as a xerophilic fungus [217].

Although the optimal temperature for growth of *P. roqueforti* is around 25 °C, it shows excellent growth kinetics at relatively low temperatures (>4 °C), so that it is also considered an altering agent of refrigerated stored foods, and grows well at temperatures typical of ripening rooms (10-13 °C) [183, 211]. In addition, *P. roqueforti* assimilates more efficiently the carbon source at 20 °C than at 25 °C [218], which casts doubt on which is actually its optimal growth temperature.

P. roqueforti tolerates well great pH variations (3-10), showing an optimal development at low pHs around 4.0-4.5 [169, 206, 211]. As the pH of the initial curd has similar values, development of the fungus is optimal in these conditions. *P. roqueforti* can also grow with lactic acid concentrations of 5%, and tolerates 0.5% of acetic acid, which is used in occasions as preservative [169, 184].

Role of *P. roqueforti* in Ripening

P. roqueforti grows in the grooves and microcavities of the cheese dough (Fig. **3**), reaching its maximum growth between 30 and 90 days [182, 184, 192, 194]. The presence of mycelium with conidiophores becomes maximum towards the central part of the cheese, where the initial concentration of salt is lowest [197, 202]. In this point, the cheese pieces are typically wrapped in foil and/or transferred to a cooler room for the rest of the ripening time [184, 197].

The use of *P. roqueforti* pursues the following objectives: (*i*) deacidification of the curd (*ii*) to generate enough proteolytic power so as to get a suitable texture (iii) to develop a high lipolytic activity that produce the flavor and taste typical of the blue cheeses.

Deacidification of the Curd:

During the initial steps of the preparation of the curd lactic acid is formed by a mixture of homolactic and heterolactic fermentations, carried out by the initial microbiota. Therefore, along with lactate, little amounts of ethanol, acetic acid and CO_2 are formed [184, 192, 201]. *P. roqueforti* can use the lactate catabolically and oxidize it to CO_2 and H_2O through the Krebs cycle, or transform it to acetate and CO_2 [182, 209]. Lactate is quickly and almost totally degraded, which causes a progressive increase of the pH, also enhanced by the proteolytic activity [182, 184]. Acetate becomes the most relevant product, given its contribution to the flavor of the cheese and the fact that it can be also generated as a product of amino acid catabolism [184, 192, 209].

Proteolytic Activity:

Proteolysis is the more complex and varied of the events occurring during the ripening of the blue cheese. It consists mainly of a biochemical process exerted on the curd caseins: α_{s1}-, α_{s2}-, β- and *para-κ*-caseins [197, 209]. As in most cheeses, the proteolytic action comply the following functions: (i) it modifies the texture of the cheese, due to the breaking up of the protein net, causing a decrease of the a_w and an increase of pH; (ii) it facilitates the release of sapid compounds during mastication, due to the modification of the protein matrix of caseins; (iii) it directly contributes to the flavor of the product due to the formation of peptides of different length and free amino acids; (iv) it releases the substrate (amino acids) for secondary catabolic processes, such as deaminations, decarboxilations, transaminations, desulfurations, aromatic acid catabolism, and reactions of free amino acids with other compounds [219, 220]. As proteins are the only continuous solid phase of the cheese, their role in the texture of the cheese is essential, producing a softer, less elastic dough [182, 192].

In blue cheeses, initial hydrolysis of caseins may occur due to the coagulant agent used, and to a lesser extent by the plasmin (the dominant indigenous milk proteinase), resulting in the formation of long and medium-sized chain peptides which are subsequently degraded by *P. roqueforti* proteases, being α_{s1}-CN(f24–199) the most abundant peptide [192, 220, 221, 222]. To provide enough proteolytic power, *P. roqueforti* possesses a complex hydrolytic system consisting of endopeptidases and exopeptidases. Endopeptidases, which are extracellularly secreted, include aspartyl protease and a metalloproteinase [83, 192, 219, 223], whereas exopeptidase activities are predominantly intracellular and consist of an acidic carboxypeptidase and an alkaline aminoprotease [192, 223, 224].

Proteolytic action is very intense in the blue cheeses and releases high as well as medium and low molecular weight peptides, along with free amino acids [219, 220]. By the end of the ripening process, *P. roqueforti* achieves almost complete exhaustion of α_{s1}-casein and hydrolization of β-casein [191, 192, 223]. The main extracellular proteolytic activity of *P. roqueforti* is due to aspartyl protease, encoded by the *aspA* gene, whose expression is induced by casein and repressed by ammonium and alkaline pH [154]. The peptides generated by hydrolysis can induce *aspA* expression even at alkaline pHs, however, alkaline conditions affect post-traductional processing of aspartyl protease, preventing the proenzyme to become mature and active [225]. Both the aspartyl protease and the metalloproteinase possess a specific activity on the α_{s1}- and β-caseins [192]. The metalloproteinase acts mostly on the βA1 and βA2-caseins at the beginning of the ripening process, but its contribution to the total proteolytic action is lower than that of the aspartyl protease, which acts on the α_{s1}-casein from the beginning of the ripening process and on the β-caseins at the end, even though the pH be not so favorable for its optimal functioning [219]. However, this behavior depends on the strain used, as it has been shown that some strains act preferentially on β-caseins from the beginning and others act equally on both α- and β-caseins [223].

On their part, exopeptidases degrade peptides present in the medium, generating high amounts of short-chain peptides and free amino acids, where they contribute to the background flavor of blue cheeses [220]. Some reports indicate that the relative proportions of individual amino acids in different types of cheese are similar [209]. In the blue cheeses, it is known that triptamine, tiramine and histamine (non-volatile amines) are present [192, 219]. Recently, Toelstede and Hofmann [226] found gamma-glutamyl peptides at high concentrations in the Blue Shropshire, which were higher than in other blue cheeses analyzed. In this variety, *P. roqueforti* produces the gamma-glutamyl peptides by a gamma-glutamyl transferase (GGT), a process in which the amino acid L-methionine seems to play a special role. This kind of peptides produces an attractive "kokumi" flavor, which is part of the organoleptic profile of many blue cheeses [226].

Figure 3: *P. roqueforti* growing inside the grooves and microcavities of a blue cheese.

Lipolytic Action:

The lipolytic activity of blue cheeses is much more pronounced than in any other type of cheese, and it is the key activity for obtaining the chemical compounds responsible for the typical flavor of these cheeses, in which their organoleptic quality lies [133]. In general terms, the lipolytic activity occurs, in first place, by the hydrolysis of triglycerides present in the cheese, generating glycerol and free fatty acids (FFA) of different length, some of which volatile. Next, the FFA are enzymatically transformed to methylketones, which in turn are reduced to secondary alcohols, events that produce all of them high quantities of diverse aromatic substances [195, 201, 227]. The FFA themselves contribute to the flavor of the product apart from functioning as precursors of methylketones [83, 209, 228].

The agent responsible for this biotransformation is *P. roqueforti*. However, in the cases in which milk is not pasteurized, native lipases may contribute to the degradation of triglycerides at the beginning of the ripening process [192, 210, 227]. On the other hand, and given that yeasts form part of the initial and accompanying microbiota [200, 202, 205, 207], they may also play a role, not only in the initial hydrolysis, but also participating in the generation of ester-type aromatic compounds [193].

The lipolytic arsenal of *P. roqueforti* has particular and interesting effects [210, 227]. It is known that this fungus produce two extracellular lipases: an acidic one and an alkaline one with an optimal pH of 9–9.5 [83, 228, 229, 230].

Both the acidic and the alkaline lipase have catalytic activity in the cheese dough [210]. As a result of the proteolytic action and the utilization of lactic acid, the pH is increased up to values close to 6 depending on the conditions of the ripening [184, 193, 194]. These pH values may undergo a later decrease when *P. roqueforti* lipases start their hydrolytic action releasing fatty acids such as butyric, caproic, caprylic and capric acids [194, 210, 228]. These processes generate the pH conditions adequate for the activity of the acidic lipase, which will then prevail upon alkaline lipase [228, 229, 231].

In the blue cheeses, lipolysis generates more than 95% of the FFA, with a high content of short-chain FFA. No other type of cheese shows so high concentrations of FFA as blue cheeses [191, 209, 228]. The lipolytic system of *P. roqueforti* permits a wide spectrum of FFA, which is noticeable in the flavor of the cheese [201, 224, 232]. When comparing the FFA profile with that of the Brie-type cheeses, it was found that in contrast to these, blue cheeses showed high medium-chain FFA content (lauric and myristic acids) [196]. In addition, *P. roqueforti* catabolizes the FFA, generating mainly 2-methylcetones (a homologous series of alkan-2-ones with odd-numbered carbon chains from C3 to C15), principally heptan-2-one, nonan-2-one and undecan-2-one, compounds that predominate in the organoleptic profile of blue cheeses [195, 228, 232, 233]. FFA may be oxidized by a thiolase in the Krebs cycle, but the fungus diverts this metabolic pathway; FFA are oxidized by β-oxidation to β-ketonic acids, and then decarboxylated to methylketones by a β-keto-acyl-decarboxylase [83, 193, 195, 232] (Fig. **4**).

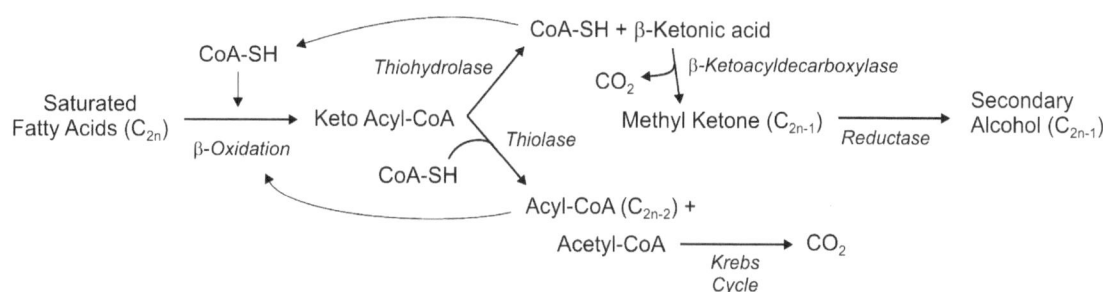

Figure 4: Catabolic pathways for degradation of fatty acids by *P. roqueforti*.

This process is important because a high concentration of medium-chain FFA may negatively affect the growth of the fungus, and thus *P. roqueforti* uses this strategy to regulate the concentration of this type of FFA [184, 227, 228]. Similarly, the enzymatic versatility of the fungus allows it to regulate the amount of methylketones, reducing them to secondary alcohols [193, 232, 233]. Both methylketones and secondary alcohols are very volatile and responsible mainly for the typical flavor of the blue cheeses [195, 233, 234]. Both products are formed in variable amounts depending on the *P. roqueforti* strain used [194, 227, 234].

Andrastins as a Possible Added Value of the Use of *P. roqueforti* as Starter Culture

Filamentous fungi possess a metabolic versatility still to be fully discovered and characterized. They produce a great variety of secondary metabolites, some of which used currently in pharmaceutical applications, such as penicillin, the most widely used antibiotic, lovastatin, an anticholesterolemic drug, or the immunosuppressant cyclosporin. One of the areas with greater interest in the pharmaceutical industry is the development of anticarcinogenic drugs [235]. In the search for new bioactive compounds, it was reported in 1996 the characterization of andrastin, a farnesyltransferase inhibitor produced by a strain of the genera *Penicillium* sp., which presents four different chemical forms, named A through D [236, 237, 238]. Almost a decade later, Nielsen *et al.* [239], reported the detection of the four types of andrastin directly on samples of European blue cheeses, all of which ripened using *P. roqueforti* as starter. In all 23 samples analyzed, andrastin A was consistently detected, with average concentrations of 2.4 µg/g, whereas types B, C and D were present at concentrations 3- to 20-fold lower than type A [239]. Later, O'Brien *et al.* [240] found that, in laboratory conditions, *P. roqueforti* consistently produced andrastin A with yields 10-fold higher than those found in blue cheeses. Production of andrastins has been also reported in other *Penicillium* species, as *P. paneum*, *P. crustosum* and *P. albocoremium* [240, 241, 242].

These results raised interest due to the potential anticarcinogenic activity of andrastins. The Ras protein is estimated to be involved in at least 30% of all cases of cancer in humans. Ras localizes anchored to the plasma membrane, in the cytosolic domain, and upon activation by membrane receptors or intermediate effectors triggers signal transduction pathways, as the MAP kinase phosphorylation cascade, which control different vital cellular processes like cell division or cytoskeleton organization [243, 244, 245]. Some mutations in the Ras encoding genes generate constitutively active forms of the protein, which will then maintain permanently active the signal transduction pathways, thus deregulating the cell cycle and predisposing the cell to become malignant [235, 243]. Ras proteins are biologically active only if they are anchored to the plasma membrane, which is achieved by the post-translational incorporation of an aliphatic isoprenoid: the farnesyl group (C_{15}), to a cysteine residue within the consensus sequence CAAX (C: cysteine; A: any aliphatic amino acid; X: serine, leucine, glutamine or methionine) situated at the C-terminal end of the protein [235, 243, 245]. This reaction is catalyzed by the enzyme farnesyltransferase, a heterodimer of 48 kDa α and 46 kDa β subunits. Mutated forms of Ras will remain inactive if farnesylation is avoided [245], and andrastins comply this function [237, 238].

The possible therapeutic use of andrastins is a very interesting possibility that must be evaluated with the usual protocols and trials. However, the possible implications of the presence of andrastins in blue cheeses raise both expectancies and concerns [239]. Production of this metabolite by *P. roqueforti* on the blue cheese may be an additional positive factor for this product; nevertheless, the effect on human health of a continuous intake of andrastins must be carefully evaluated. The questions that must be taken into account can be summarized as follows: (i) is the amount of andrastins naturally present in cheese sufficient to induce any kind of positive or negative effect?, (ii) could the steady consumption of blue cheese protect from the appearance of malignant tumors?, (iii) what is the real effect of a continuous intake of andrastins in small doses on healthy individuals?. Thorough studies to answer these questions must be carried out, and only positive results would permit to classify blue cheeses among functional foods. On the contrary, if results turn out to be negative, elaboration of blue cheeses should be modified, for instance modifying genetically *P. roqueforti* strains so that they do not produce andrastins.

ACKNOWLEDGEMENTS

Supported by grants Fondecyt 7070044 (to F.F. and R.C.), Fondecyt 11060003 and DICYT-USACH (to R.C).

REFERENCES

[1]   Rolfe T, Rolfe FW. The Romance of the Fungus World. Philadelphia: Lipincott 1928.

[2]   Ramsbottom J. Mushrooms and Toadstools. London: Collins 1953.

[3]   Allegro JM. The Sacred Mushroom and the Cross. London: Hodder & Stoughton 1970.

[4]   Wang YC. Mycology in China with emphasis on review of the ancient literature. Acta Mycol Sin 1985; 4: 133-40.

[5]   Willis JH. A bibliography of 'Blackfellows Bread', *Polyporus mylittae* Cooke & Massee. Muelleria 1967; 1: 203-12.

[6]   Weber GF. The occurrence of tuckahoes and *Poria cocos* in Florida. Mycologia 1929; 21: 113-30.

[7]   Hawksworthd L, Sutton BC, Ainsworth GC. Ainsworth and Bisby's Dictionary of the Fungi 7th ed. Commonwealth Agricultural Bureaux: Commonwealth Mycological Institute 1983.

[8]   Campbell-Platt G, Cook PE. Fungi in the production of foods and food ingredients. J Appl Bacteriol Symp Suppl 1989; 67: 117S-31S.

[9]   Deacon JW. Modern Mycology, 3rd ed. Blackwell Science, USA 1997.

[10]   Edwards G. Myco-protein- The development of a new food. Food Laboratory Newsletter May 1986, pp. 21-4.

[11]   Hesseltine CW. A millennium of fungi, food and fermentation. Mycologia 1965; 57: 149-97.

[12]   Hesseltine CW. Microbiology of oriental fermented foods. Annu Rev Microbiol 1983; 37: 571-601.

[13]   Yokotsuka T. Fermented protein foods in the Orient, with emphasis on shoyu and miso in Japan. In: Wood BJB Ed. Microbiology of fermented foods. London, New York: Elsevier Applied Science Publishers 1985; pp. 197-246.

[14]   Ko SD. Fermentation of foods by moulds. In: Samson RA, Hoekstra ES, Van Oorschot CAN, Eds. Introduction to food-borne fungi. Delft: Centralbureau voor Schimmelcultures 1981; pp. 326-41.

[15]   Wolf G. Traditional fermented food. In: Anke T. Ed. Fungal Biotechnology. Weinheim: Chapman & Hall 1997; pp. 3-13.

[16]   Mintzlaff HJ, Leistner L. Untersuchungen Zur Selektion eines technologish geeigneten und toxikologisch unbodenklichen Schimmelpilz-Stammes fur die Rohwurst-Herstellung. Zbl Vet Med B 1972; 19: 291-300.

[17]   Hwang HJ, Vogel RF, Hammes WP. Entwicklung von Schimmelpilzkulturen für die Rohwurstherstellung. Charakterisierung der Stämme und toxikologische Bewertung. Fleischwirtschaft 1993; 73: 89-92.

[18] Hwang HJ, Vogel RF, Hammes WP. Entwicklung von Schimmelpilzkulturen für die Rohwurstherstellung. Technologische Eignung der Stämme und sensorische Bewertung der Produkte. Fleischwirtschaft 1993; 73: 327-32.

[19] Farnworth ER. Ed. Handbook of fermented functional foods, 2nd ed. Boca Ratón: CRC Press.

[20] Frisvad JC, Thrane U. Mycotoxin production by common filamentous fungi. In: Samson RA Hoekstra ES Frisvad JC Filtenborg O, Eds. Introduction to Food and Airborne Fungi 6 ed. Centraalbureau voor Schimmelcultures Utrecht 2002; pp. 321-31.

[21] Papagianni M, Ambrosiadis I, Filiousis G. Mould growth on traditional Greek sausages and penicillin production by *Penicillium* isolates. Meat Sci 2007; 76: 653-7.

[22] Filtenborg O, Frisvad JC, Thrane U. Moulds in food spoilage. Int J Food Microbiol 1996; 33: 85-102.

[23] Laich F, Fierro F, Martín JF. Production of penicillin by fungi growing on food products: Identification of a complete penicillin gene cluster in *Penicillium griseofulvum* and a truncated cluster in *Penicillium verrucosum*. Appl Environ Microbiol 2002; 68: 1211-9.

[24] Leistner L, Eckardt C. Vorkommen toxinogener Penicillien bei Fleischerzeugnissen. Fleischwirtschaft 1979; 59: 1892-6.

[25] Pitt JI. Toxigenic *Penicillium* species. In: Doyle MP, Beuchat LR, Montville TJ, Eds. Food Microbiology. Fundamentals and Frontiers. Washington: ASM Press 1997; pp. 406-18.

[26] Frisvad JC, Smedsgaard J, Larsen TO, Samson RA. Mycotoxins drugs and other extrolites produced by species in *Penicillium* subgenus *Penicillium*. Stud Mycol 2004; 49: 201-41.

[27] Pestka J. Fungal toxins in raw and fermented meats. In: Campbell-Platt G, Cook PE, Ed. Fermented meats. London: Blackie Academic and Professional 1995; pp. 194-216.

[28] Laich F, Fierro F, Cardoza RE, Martín JF. Organization of the gene cluster for biosynthesis of penicillin in *Penicillium nalgiovense* and antibiotic production in cured dry sausages. Appl Environ Microbiol 1999; 65: 1236-40.

[29] Geisen R, Lücke FK, Kröckel L. Starter and protective cultures for meat and meat products. Fleischwirtschaft 1992; 72: 894-8.

[30] Singh BJ, Dincho D. Molds as protective cultures for raw dry sausages. J Food Prot 1994; 57: 928-30.

[31] Laich F, Fierro F, Martín JF. Isolation of *Penicillium nalgiovense* strains impaired in penicillin production by disruption of the *pcbAB* gene and application as starters on cured meat products. Mycol Res 2003; 107: 717-26.

[32] Geisen R, Glenn E, Leistner L. Two *Penicillium camemberti* mutants affected in the production of cyclopiazonic acid. Appl Environ Microbiol 1990; 56: 3587-90.

[33] Frisvad JC, Samson RA. Filamentous fungi in foods and feeds. Ecology, spoilage and mycotoxin production. In: Arora DK, Mukerji KG, Marth EH, Eds. Handbook of Applied Mycology New York: Marcel Dekker 1991; vol. 3. pp. 31-68.

[34] Frank JK, Orth R, Ivankowic S, Kuhlmann M, Schmähl D. Investigations on carcinogenic effects of *Penicillium caseicolum* and *P. roquefoti* in rats. Cell Mol Life Sci 1977; 33: 515-6.

[35] Geisen R. Fungal starter cultures for fermented foods: molecular aspects. Trend Food Sci Technol 1993; 4: 251-6.

[36] Fierro F, Laich F, García-Rico RO, Martín JF. High efficiency transformation of *Penicillium nalgiovense* with integrative and autonomously replicating plasmids. Int J Food Microbiol 2004; 90: 237-48.

[37] Geisen R, Ständner L, Leistner L. New mould starter cultures by genetic modification. Food Biotechnol 1990; 4: 497-504.

[38] Geisen R. Expression of the *Aspergillus niger* glucose oxidase gene in *Penicillium nalgiovense*. World J Microbiol Biotechnol 1995; 11: 322-5.

[39] Geisen R. Inhibition of food-related pathogenic bacteria by god-transformed *Penicillium nalgiovense* strains. J Food Prot 1999; 62: 940-3.

[40] Bok JW, Keller NP. LaeA, a Regulator of Secondary Metabolism in *Aspergillus* spp. Eukaryot Cell 2004; 3: 527-35.

[41] Bok JW, Noordermeer D, Kale SP, Keller NP. Secondary metabolic gene cluster silencing in *Aspergillus nidulans*. Mol Microbiol 2006; 61: 1636-45.

[42] Bok JW, Hoffmeister D, Maggio-Hall LA, Murillo R, Glasner JD, Keller NP. Genomic mining for *Aspergillus* natural products. Chem Biol 2006; 13: 31-7.

[43] Perrin RM, Fedorova ND, Bok JW, *et al.* Transcriptional regulation of chemical diversity in *Aspergillus fumigatus* by LaeA. PLoS Pathog 2007; 3: 0508-17.

[44] Kosalkova K, García-Estrada C, Ullán RV, *et al.* The global regulator LaeA controls penicillin biosynthesis, pigmentation and sporulation, but not roquefortine C synthesis in *Penicillium chrysogenum*. Biochimie 2009; 91: 214-25.

[45] Sugui JA, Pardo J, Chang YC, *et al.* Role of *laeA* in the regulation of *alb1*, *gliP*, conidial morphology, and virulence in *Aspergillus fumigatus*. Eukaryot cell 2007; 6: 1552-61.

[46] Campbell-Platt G. Fermented Foods of the World. A Dictionary and Guide. London: Butterworths 1987.

[47] Wood BJB. Ed. Microbiology of fermented foods, London, New York: Elsevier Applied Science Publishers 1985; vol. 1.

[48] Steinkraus KH. Handbook of indigenous fermented foods. New York: Marcel Dekker 1983.

[49]   Raper KB, Thom C. Eds. A manual of the Penicillia. Baltimore Maryland: Williams and Wilkins 1949.

[50]   Pitt JI. The genus *Penicillium* and its teleomorphic states *Eupenicillium* and *Talaromyces*. London. UK: Academic Press Inc. 1979.

[51]   Fink-Gremmels J, El-Banna A, Leistner L. Developing mould starter cultures for meat products. Fleischwirtschaft 1988; 68: 1292-4.

[52]   Andersen SJ. Taxonomy of *Penicillium nalgiovense* isolates from mould-fermented sausages. Antonie Van Leeuwenhoek 1995; 68: 165-71.

[53]   Geisen R. Characterization of the species *Penicillium nalgiovense* by RAPD and protein patterns and its comparison with *Penicillium chrysogenum*. Syst Appl Microbiol 1995; 18: 595-601.

[54]   Dupont J, Magnin S, Marti A. Molecular tools for identification of *Penicillium* starter cultures used in the food industry. Int J Food Microbiol 1999; 49: 109-18.

[55]   Raistrick H, Ziffer J. The colouring matters of *Penicillium nalgiovensis* Laxa. Biochem J 1951; 49: 563-74.

[56]   Birch AJ, Massy-Westropp RA. Studies in relation to biosynthesis. Part XI. The structure of nalgiovensin. J Chem Soc 1957; 1957: 2215-7.

[57]   Birch AJ, Stapleford KSJ. The structure of nalgiolaxin. J Chem Soc 1967; 1967: 2570-1.

[58]   Färber P, Geisen R. Antagonistic activity of the food-related filamentous fungus *Penicillium nalgiovense* by the production of penicillin. Appl Environ Microbiol 1994; 60: 3401-4.

[59]   Andersen SJ, Frisvad JC. Penicillin production by *Penicillium nalgiovense*. Lett Appl Microbiol 1994; 19: 486-8.

[60]   Kosalková K, Rodríguez-Sáiz M, Barredo JL, Martín JF. Binding of the PTA1 transcriptional activator to the divergent promoter region of the first two genes of the penicillin pathway in different *Penicillium* species. Curr Genet 2007; 52: 229-37.

[61]   Färber P, Geisen R. Kariotipe of *Penicillium nalgiovense* and assignment of the penicillin biosynthetic genes to chromosome IV. Int J Food Microbiol 2000; 58: 59-63.

[62]   Fierro F, Gutiérrez S, Díez B, Martín JF. Resolution of four large chromosomes in penicillin-producing filamentous fungi: the penicillin gene cluster is located on chromosome II (9.6 Mb) in *Penicillium notatum* and chromosome I (10.4 Mb) in *Penicillium chrysogenum*. Mol Gen Genet 1993; 241: 573-8.

[63]   Geisen R. *P. nalgiovense* carries a gene which is homologous to the *paf* gene of *P. chrysogenum* which codes for an antifungal peptide. Int J Food Microbiol 2000; 62: 95-101.

[64]   Iacumin L, Chiesa L, Boscolo D, *et al.* Moulds and ochratoxin A on surfaces of artisanal and industrial dry sausages. Food Microbiol 2009; 26: 65-70.

[65]   Bogs C, Battilani P, Geisen R. Development of a molecular detection and differentiation system for ochratoxin A producing *Penicillium* species and its application to analyse the occurrence of *Penicillium nordicum* in cured meats. Int J Food Microbiol 2006; 107: 39-47.

[66]   Pitt JI, Hocking AD. Fungi and food spoilage 3rd ed. Springer, Dordrecht, 2009.

[67]   Lücke FK. Procesos microbiológicos en la elaboración de embutidos secos y jamones crudos. Fleischwirtschaft (Spanish) 1987; 2: 39-46.

[68]   Grazia L, Romano P, Bagni D, Roggiani D, Guglielmi G. The role of moulds in the ripening process of salami. Food Microbiol 1986; 3: 19-25.

[69]   Huerta T, Sanchis V, Hernández J, Hernández E. Mycoflora of dry-salted Spanish ham. Microbiol Alim Nutr 1987; 5: 247-52.

[70]   Trigueros G, García ML, Casas C, Ordóñez JA, Selgas MD. Proteolytic and lipolytic activities of mold strains isolated from Spanish dry-fermented sausages. Z. Lebensm Unters Forch 1995; 201: 298-303.

[71]   Núñez F, Rodríguez ME, Bermúdez ME, Córdoba JJ, Asensio MA. Composition and toxigenic potential of mould population on dry-cured Iberian ham. Int J Food Microbiol 1996; 32: 185-97.

[72]   Toledo VM, Selgas MD, Casas MC, Ordóñez JA, García ML. Effect of selected mold strains on the proteolysis in Dry-fermented sausages. Z. Lebensm Unters Forch 1997; 204: 385-40.

[73]   Rodríguez M, Núñez F, Córdoba JJ, Bermúdez ME, Asensio MA. Evaluation of proteolytic activity of micro-organisms isolated from dry cured ham. J Appl Microbiol 1998; 85: 905-12.

[74]   Bruna JM, Hierro EM, de la Hoz L, Mottram DS, Fernández M, Ordoñez JA. The contribution of *Penicillium aurantiogriseum* to the volatile composition and sensory quality of dry fermented sausages. Meat Sci 2001; 59: 97-107.

[75]   Martín A, Asensio MA, Bermúdez ME, Córdoba MG, Aranda E, Córdoba JJ. Proteolytic activity of *Penicillium chrysogenum* and *Debaryomyces hansenii* during controled ripening of pork loins. Meat Sci 2002; 62: 129-37.

[76]   Martín A, Córdoba JJ, Benito MJ, Aranda E, Asensio MA. Effect of *Penicillium chrysogenum* and *Debaryomyces hansenii* on the volatile compounds during controlled ripening of pork loins. Int J Food Microbiol 2003; 84: 327-38.

[77]   Martín A, Córdoba JJ, Aranda E, Córdoba MG, Asensio MA. Contribution of a selected fungal population to the volatile compounds on dry-cured ham. Int J Food Microbiol 2006; 110: 8-18.

[78]   Selgas DM, Casas C, Toledo VM, Garcia ML. Effect of selected mould strains on lipolysis in dry fermented sausages. Eur Food Res Technol 1999; 209: 360-5.

[79]   Benito MJ, Córdoba JJ, Alonso M, Asensio MA, Nunez F. Hydrolytic activity of *Penicillium chrysogenum* Pg222 on pork myofibrillar proteins. Int J Food Microbiol 2003; 89: 155-61.

[80]   Benito MJ, Núñez F, Córdoba MG, Martín A, Córdoba JJ. Generation of nonprotein nitrogen and volatile compounds by *Penicillium chrysogenum* Pg222 activity on pork myofibrillar proteins. Food Microbiol 2005; 22: 513-9.

[81]   Ludemann V, Pose G, Pollio ML, Segura J. Determination of growth characteristics and lipolytic and proteolytic activities of *Penicillium* strains isolated from Argentinean salami. Int J Food Microbiol 2004; 96: 13-8.

[82]   Tabuc C, Bailly JD, Bailly S, Querin A, Guerre P. Toxigenic potential of fungal mycoflora isolated from dry cured meat products: preliminary study. Rev Med Vet 2004; 156: 287-91.

[83]   Kinsella JE, Hwang DH. Enzymes of *Penicillium roqueforti* involved in the biosynthesis of cheese flavor. CRC Crit Rev Food Sci Nutr 1976; 8: 191-228.

[84]   Martín A, Córdoba JJ, Núñez F, Benito MJ, Asensio MA. Contribution of a selected fungal population to proteolysis on dry-cured ham. Int J Food Microbiol 2004; 94: 55-66.

[85]   Carbone D. Descrizione di alcuni eumiceti provenienti da carni insaccate sane. Atti dell'Istituto Botanico della Università di Pavia 1910; 14: 259-325.

[86]   Leistner L, Ayres JC. Molds and meats. Fleischwirtschaft 1968; 48: 62-5.

[87]   Sutic MJ, Ayres C, Koehler PE. Identification and aflatoxin production of molds isolated from country cured hams. Appl Microbiol 1972; 23: 656-8.

[88]   Ciegler AH, Mintzlaff J, Machnik W, Leistner L. Untersuchungen über das Toxinbildungsvermögen von Rohwürsten isolierter Schimmelpilze der Gattung *Penicillium*. Fleischwirtschaft 1972; 10: 1311-8.

[89]   Wu MT. Ayres JC. Koehler E. Toxigenic Aspergilli and Penicillia isolated from aged cured meats. Appl Environ Microbiol 1974; 28: 1094-6.

[90]   Van der Riet WB. Studies on the mycoflora of biltong. S Afr Food Rev 1976; 3: 105-11.

[91]   Van der Riet WB. Biltong a South African dried meat product. Fleischwirtschaft 1982; 62: 1000-1.

[92]   Dragoni I, Marino C. Description and classification of *Penicillium* species isolates from raw ripened sausages. Arch Vet Ital 1979; 30: 142-76.

[93]   Dragoni I, Cantoni C. Le muffe negli insaccati crudi stagionati. Ind Alim (Italy) 1979; 18: 281-4.

[94]   Dragoni I, Marino C, Cantoni C. Muffe in prodotti carnei salati e stagionati (bresaole e prosciutti crudi). Ind Aliment 1980; 19: 405-7.

[95]   Dragoni I, Ravenna R, Marino C. Descrizione e classificazione delle specie di *Aspergillus* isolate dalla superficie di prosciutti stagionati di Parma e San Danielle. Arch Vet Ital 1980; 31: 1-56.

[96]   Dragoni I, Cantoni C, Aubert A. Grey mould in raw seasoned sausages. Ind Aliment 1988; 27: 976-9.

[97]   Dragoni I, Marino C, Cantoni C. Moulds on smoke bacon. Ind Aliment 1988; 27: 353-5.

[98]   Dragoni I, Cantoni C, Papa A. Typical fungal flora of hourse "bresaola" produced in Valtellina. Ind Aliment 1990; 29: 560-3.

[99]   Draughon F, Melton C, Maxedon D. Microbial profiles of country-cured hams aged in Stockinettes Barrier Bags and Paraffin wax. Appl Environ Microbiol 1981; 41: 1078-80.

[100]   Monte E, Villanueva R, Dominguez A. Fungal profiles of Spanish country-cured hams. Int J Food Microbiol 1986; 3: 355-9.

[101]   Huerta T, Sanchis V, Hernández J, Hernández E. Enzymatic activities and antimicrobial effects of *Aspergillus* and *Penicillium* strains isolated from Spanish dry cured hams: quantitative and qualitative aspects. Microbiol Alim Nutr 1987; 5: 289-94.

[102]   Philipp S, Pedersen PD. Mould cultures for the industry. A short review with special reference to the cheese and sausage production. Dan Dairy Food Ind-Worldwide 1988; 6: 8-12.

[103]   Leistner L. Mould-fermented foods: recent developments. Food Biotechnol 1990; 4: 433-41.

[104]   Motilva Casado MJ, Díaz Borrás MA, Vila Aguilar RV. Fungal flora present on the surface of cured Spanish ham. Methodological study for its isolation and identification. Fleischwirtschaft 1991; 71: 1300-2.

[105]   Kaur H, Joshi DV, Kwatra M S, Jand SK. Preliminary studies on Mycoflora of pig sausages. Ind J Comp Microbiol, Immunol Infect Dis 1992; 13: 31–3.

[106]   Rojas FJ, Jodral M, Gosalvez F, Pozo R. Mycoflora and toxigenic *Aspergillus flavus* in Spanish dry-cured ham. Int J Food Microbiol 1991; 13: 249-55.

[107]   Andersen SJ. Compositional changes in surface mycoflora during ripening of naturally fermented sausages. J Food Prot 1995; 58: 426-9.

[108] Peintner U, Geiger J, Pöder R. The mycobiota of Speck a traditional Tyrolean smoked and cured ham. J Food Prot 2000; 63: 1399-403.

[109] López-Diaz TM, Santos JA, Garcia-Lopez ML, Otero A. Surface mycoflora of a Spanish fermented meat sausage and toxigenicity of *Penicillium* isolates. Int J Food Microbiol 2001; 68: 69-74.

[110] Mizakova A, Pipova M, Turek P, The occurrence of moulds in fermented raw meat products. Czech J Food Sci 2002; 20: 89-94.

[111] Comi G, Orlic S, Redzepovic S, Urso R, Iacumin L. Moulds isolated from Istrian dried ham at the pre-ripening and ripening level. Int J Food Microbiol 2004; 96: 29-34.

[112] Wang X, Ma P, Jiang D, Peng Q, Yang H. The natural microflora of Xuanwei ham and the no-mouldy ham production. J Food Eng 2006; 77: 103-11.

[113] Battilani P, Pietri VA, Giorni P, *et al.* *Penicillium* populations in dry-cured ham manufacturing plants. J Food Prot 2007; 70: 975-80.

[114] Sorensen LM, Jacobsen T, Nielsen PV, Frisvad JC, Koch AG. Mycobiota in the processing areas of two different meat products. Int J Food Microbiol 2008; 124: 58-64.

[115] Sunesen LO, Stahnke LH. Mould starter cultures for dry sausages - selection application and effects. Meat Sci 2003; 65: 935-48.

[116] Asefa DT, Gjerde RO, Sidhu MS, Langsrud S, Kure CF, Nesbakken T, Skaar I. Moulds contaminants on Norwegian dry-cured meat products. Int J Food Microbiol 2009; 128: 435-9.

[117] Cesari EP. La maturation du saucisson. Acad Sci Paris 1919; 168: 802.

[118] Cesari EP, Guillermond A. Les levures des saucisson. Ann Inst Pasteur 1920; 34: 229.

[119] Castellari C Molina Favero C Quadrelli AM Laich F. Effect of temperature, water activity, pH and NaCl in *P. nalgiovense* and *P. chrysogenum* conidiogenesis. Proceedings of the XVII Congreso Latinoamericano y X Argentino de Microbiología. Buenos Aires: Argentina 2004; pp. 253.

[120] Lücke FK, Hechelmann H. Cultivos starter para embutido seco y jamón crudo. Composición y efecto. Fleischwirtschaft (Spanish) 1988; 1: 38-48.

[121] Castellari C, Oliverio G, Buenader S, Quadrelli AM, Laich F. Inoculation of dry meat sausages with native strains of *Penicillium nalgiovense*. Proceedings of the III International Food Science and Technology Congress. Córdoba, Argentina 2006; pp. 540.

[122] Horwitz C, Wehner FC. Antibiotics in mould-cured salami. S Afr Med 1977; 52: 669.

[123] Kanny G, Puygrenier J, Beaudoin E, Moneret-Vautrin DA. Alimentary anaphylactic shock: implication of penicillin residues. Allerg Immunol 1994; 26: 181-3.

[124] Lindemayr HR, Knobler D, Kraft Baumgartner W. Challenge of penicillin-allergic volunteers with penicillin-contaminated meat. Allergy 1981; 36: 471-8.

[125] Shoemaker NB, Vlamakis H, Hayes K, Salyers AA. Evidence for extensive resistance gene transfer among *Bacteroides* spp. and among *Bacteroides* and other genera in the human colon. Appl Environ Microbiol 2001; 67: 561-8.

[126] Heinrich P, Rosenstein R, Bohmer M, Sonner P, Götz F. The molecular organization of the lysostaphin gene and its sequences repeated in tandem. Mol Gen Genct 1987; 209: 563-9.

[127] Kriechbaum M, Heilmann J, Wientjes FJ, *et al.* Cloning and DNA sequence analysis of the glucose oxidase gene from *Aspergillus nidulans* NRRL-3. FEBS Lett 1989; 255: 63-6.

[128] Quilleauquy M, Castellari C, Oliverio G, Buenader S, Laich F. *Penicillium nalgiovense* viability in modified atmosphere packaging of dry fermented sausages. Proceedings of the VI Congreso Latinoamericano de Micología. Mar del Plata, Argentina 2008; pp. 209.

[129] Bruna, JM, Fernández M, Ordóñez JA, de la Hoz L. Enhancement of the flavor development of dry fermented sausages by using a protease (pronase E) and a cell-free extract of *Penicillium camemberti*. J Sci Food Agric 2002; 82: 526-33.

[130] Samson RA, Eckardt C, Orth R. The taxonomy of *Penicillium* species from fermented cheeses. Antonie van Leeuwenhoek 1977; 43: 341-50.

[131] Cerning J, Gripon JC, Lamberet G, Lenoir J. Les activities biochimiques des *Penicillium* utilisés en fromagerie. Lait 1987; 67: 3-39.

[132] Fuke Y, Matsuoka H. The purification and characterization of prolyl aminopeptidase from *Penicillium camemberti*. J Dairy Res 1993; 76: 2478-84.

[133] McSweeney PLH, Sousa MJ. Biochemical pathways for the production of flavour compounds during ripening: a review. Lait 2000; 80: 293-324.

[134] Karahadian C, Josephson DB, Lindsay RC. Contribution of *Penicillium* sp. to the flavors of brie and camembert cheese. J Dairy Res 1985; 68: 1865-77.

[135] Bruna, JM, Hierro EM, de la Hoz L, Mottram DS, Fernández M, Ordóñez JA. Changes in selected biochemical and sensory parameters as affected by the superficial inoculation of *Penicillium camemberti* on dry fermented sausages. Int J Food Microbiol 2003; 85: 111-25.

[136] Dox, AW. The intracellular enzymes of lower fungi, especially those of *Penicillium camemberti*. J Biol Chem 1909; 6: 461-7.

[137] Gripon JC. Flavor and texture in soft cheeses. In: Law BA, Ed. Microbiology and biochemistry of cheese and fermented milk. London: Bleckie Academia & Professional 1997; pp. 193-206.

[138] Upadhyay VK, McSweeney PLH. Acceleration of cheese ripening. In: Smit G, Ed. Dairy processing: improving quality. Cambridge: Woodhead Publishing 2003; pp. 419-447.

[139] Yamaguchi S, Mase T. Purification and characterization of mono- and diacylglycerol lipase isolated from *Penicillium camembertii* U-150. Appl Microbiol Biotechnol 1991; 34: 720-5.

[140] Isobe K, Nokihara K, Yamaguchi S, Mase T, Schmid RD. Crystallization and characterization of monoacylglycerol and diacylglycerol lipase from *Penicillium camemberti*. Eur J Biochem 1992; 203: 233-7.

[141] Isobe K, Nokihara K. Primary structure determination of mono- and diacylglycerol lipase from *Penicillium camemberti*. FEBS Lett 1993; 320: 101-6.

[142] Yamaguchi S, Mase T, Takeuchi K. Secretion of mono- and diacylglycerol lipase from *Penicillium camembertii* U-150 by *Saccharomyces cerevisiae* and site-directed mutagenesis of the putative catalytic sites of the lipase. Biosci Biotech Biochem 1992; 56: 315-9.

[143] Isobe K, Aumann KD, Schmid RD. A structural model of mono- and diacylglycerol lipase from *Penicillium camembertii*. J Biotechnol 1994; 32: 83-8.

[144] Larsen MD, Jensen K. The effects of environmental conditions on the lipolytic activity of strains of *Penicillium roqueforti*. Int J Food Microbiol 1999; 46: 159-66.

[145] Ruiz B, Farrés A, Langley E, Masso F, Sánchez S. Purification and characterization of an extracellular lipase from *Penicillium candidum*. Lipids 2001; 36: 283-9.

[146] Okumura J, Kinsella JE. Methyl ketone formation by *Penicillium camemberti* in model systems. J Dairy Sci 1985; 68: 11-5.

[147] Jollivet N, Belin J-M. Comparison of volatile flavor compounds produced by ten strains of *Penicillium camemberti* Thom. J Dairy Sci 1992; 76: 1837-44.

[148] Kermasha S, Perraud X, Bisakowski B, Husson F. Production of flavor compounds by hydroperoxide lyase from enzymatic extracts of *Penicillium* sp. J Mol Catal B 2002; 19-20: 479-87.

[149] Husson F, Krumov KN, Cases E, Cayot P, Bisakowski B, Kermasha S, Belin J-M. Influence of medium composition and structure on the biosynthesis of the natural flavour 1-octen-3-ol by *Penicillium camemberti*. Process Biochem 2005; 40: 1395-400.

[150] Husson F, Thomas M, Kermasha S, Belin J-M. Effect of linoleic acid induction on the production of 1-octen-3-ol by the lipoxygenase and hydroperoxide lyase activities of *Penicillium camemberti*. J Mol Catal B 2002; 19-20: 363-9.

[151] Chrzanowska J, Kolaczkowska M, Dryjanski M, Stachowiak D, Polanowski A. Aspartic proteinase from *Penicillium camemberti*: purification, properties, and substrate specificity. Enzyme Microbiol Technol 1995; 17: 719-24.

[152] Gripon JC, Auberger B, Lenoir J. Metalloproteases from *Penicillium caseicolum* and *Penicillium roqueforti*. Comparison of specificity and chemical characterization. Int J Biochem 1980; 12: 451-5.

[153] Lenoir J, Auberger B. Les caracteres du systeme proteolytique de *Penicillium caseicolum* II. Caracterisation d'une protease neutre. Lait 1977; 57: 471-91.

[154] Gente S, Durand-Poussereau N, Fevre M. Control of the expression of *aspA*, the aspartyl protease gene from *Penicillium roqueforti*. Mol Gen Genet 1997; 256: 557-65.

[155] Desmazeaud MJ, Gripon J-C, Le Bars D, Bergere JL. Etude du role des micro-organismes et des enzymes au cours de la maturation des fromages. Lait 1976; 557: 379-96.

[156] Gripon JC, Desmazeaud MJ, Le Bars D, Bergere J-L. Role of proteolytic enzymes of *Streptococcus lactis*, *Penicillium roqueforti* and *Penicillium caseicolum* during cheese ripening. J Dairy Sci 1977; 60: 1532-8.

[157] Trieu-Cuot P, Archieri-Haze M-J, Gripon JC. Etude comparative de l'action des métalloprotéases de *Penicillium caseicolum* et *Penicillium roqueforti* sur les caséines alpha$_{S1}$ et beta. Lait 1982; 62: 234-49.

[158] Grapin R, Rank TC, Olson NF. Primary proteolysis of cheese proteins during ripening. A review. J Dairy Sci 1985; 68: 531-40.

[159] Hjört CM. Production of food additives using filamentous fungi. In: Heller KJ, Ed. Genetically engineered food. KGaA Weinheim: Wiley-VCH Verlag GmbH&Co. 2003; pp. 87-99.

[160] Faus I, Patiño C, del Río JL, *et al.* Expression of a synthetic gene encoding the sweet-tasting protein thaumatin in the filamentous fungus *Penicillium roquefortii*. Biotechnol Lett 1997; 19: 1185-91.

[161] Durand R, Rascle C, Fèvre M. Expression of a catalytic domain of a *Neocallimastix frontalis* endoxylanase gene (*xyn3*) in *Kluyveromyces lactis* and *Penicillium roqueforti*. Appl Microbiol Biotechnol 1999; 52: 208-14.

[162] Washizu K, Yamaguchi S, inventors. Plasmid for expressing secretion of polypeptide usable in mold and yeast and production of polypeptide using the same. Japanese patent N° JP7123987. 1995.

[163] Jaeger KE, Eggert T. Lipases for biotechnology. Curr Opin Biotechnol 2002; 13: 390-7.

[164] Tamaguchi S, Mase T. High-yield synthesis of monoglyceride by mono- and diacylglycerol lipase from *Penicillium camembertii* U-150. J Ferment Bioeng 1991; 72: 162-7.

[165] Abraham WR, Arfmann HA. *Penicillium camemberti* a new source of brefeldin A. Planta Med 1992; 58: 484.

[166] Taseli BK, Gökcay CF, Taseli H. Upflow column reactor design for dechlorination of chlorinated pulping wastes by *Penicillium camemberti*. J Environ Manage 2004; 72: 175-9.

[167] Taseli BK, Gökcay CF. Degradation of low and high molecular weight fractions of softwood bleachery effluents by *Penicillium camemberti* in up-flow column reactor. Bull Environ Contam Toxicol 2006; 76: 481-9.

[168] Lund F, Filtenborg O, Westall S, Frisvad JC. Associated mycoflora of rye bread. Lett Appl Microbiol 1996; 23: 213-7.

[169] Pitt JI, Hocking AD. Fungi and food spoilage. 2nd ed. Cambridge: Blackie Academic & Professional 1997.

[170] Fujikawa H, Wauke T, Kusunoki J, *et al.* Contamination of microbial foreign bodies in bottled mineral water in Tokyo, Japan. J Appl Microbiol 1997; 82: 287-91.

[171] Kure CF, Wasteson Y, Brendehaug J, Skaar I. Mould contaminants on Jarlsberg and Norvegia cheese blocks from four factories. Int J Food Microbiol 2001; 70: 21-7.

[172] Sumarah MW, Miller JD, Blackwell BA. Isolation and metabolite production by *Penicillium roqueforti*, *P. paneum* and *P. crustosum* isolated in Canada. Mycopathologia 2005; 159: 571-7.

[173] Nielsen KF, Sumarah MW, Frisvad JC, Miller JD. Production of metabolites from the *Penicillium roqueforti* complex. J Agric Food Chem 2006; 54: 3756-63.

[174] Mann DA, Beuchat LR. Combinations of antimycotics to inhibit the growth of molds capable of producing 1,3-pentadiene. Food Microbiol 2008; 25: 144-53.

[175] O'brien M, Egan D, O'kiely P, Forristal PD, Doohan FM, Fuller HT. Morphological and molecular characterisation of *Penicillium roqueforti* and *P. paneum* isolated from baled grass silage. Mycol Res 2008; 112: 921-32.

[176] Valerio F, Favilla M, De Bellis P, Sisto A, de Candia S, Lavermicocca P. Antifungal activity of strains of lactic acid bacteria isolated from a semolina ecosystem against *Penicillium roqueforti*, *Aspergillus niger* and *Endomyces fibuliger* contaminating bakery products. Syst Appl Microbiol 2009; 32: 438-48.

[177] Valík L, Baranyi J, Görner F. Predicting fungal growth: the effect of water activity on *Penicillium roqueforti*. Int J Food Microbiol 1999; 47: 141-6.

[178] Kure CF, Skaar I, Brendehaug J. Mould contamination in production of semi-hard cheese. Int J Food Microbiol 2004; 93: 41-9.

[179] Decker M, Nielsen PV. The inhibitory effect of *Penicillium camemberti* and *Geotruchum candidum* on the associated funga of white mould cheese. Int J Food Microbiol 2005; 104: 51-60.

[180] Hayaloglu AA, Kirbag S. Microbial quality and presence of moulds in Kuflu cheese. Int J Food Microbiol 2007; 115: 376-80.

[181] Meyers E, Knight SG. Studies on the nutrition of *Penicillium roqueforti*. Appl Microbiol 1958; 6: 174-8.

[182] Coghill D. The ripening of blue vein cheese: a review. Aust J Dairy Technol 1979; 34: 72–5.

[183] López-Díaz TM, Santos J, Otero A, Garcia ML, Moreno B. Some technological properties of *Penicillium roqueforti* strains isolated from a home-made blue cheese. Lett Appl Microbiol 1996; 23: 5-8.

[184] Farkie NY, Vedamuthu ER. In: Robinson RK, Ed. Dairy Microbiology Handbook: The Microbiology of Milk and Milk Products. 3rd ed. New York: John Wiley and Sons, Inc. 2002; pp. 479-510.

[185] O'Brien M, Nielsen KF, O'Kiely P, Forristal PD, Fuller HT, Frisvad JC. Mycotoxins and other secondary metabolites produced *in vitro* by *Penicillium paneum* Frisvad and *Penicillium roqueforti* Thom isolated from baled grass silage in Ireland. J Agric Food Chem 2006; 54: 9268-76.

[186] López-Díaz TM, Román-Blanco C, García-Arias MT, García-Fernández MC, García-López ML. Mycotoxins in two Spanish cheese varieties. Int J Food Microbiol 1996; 30: 391-5.

[187] Finoli C, Vecchio A, Galli A, Dragoni I. Roquefortine C occurrence in blue cheese. J Food Prot 2001; 64: 246-51.

[188] Erdogan A, Sert S. Mycotoxin-forming ability of two *Penicillium roqueforti* strains in blue moldy tulum cheese ripened at various temperatures. J Food Prot 2004; 67: 533-5.

[189] Taniwaki MH, Hocking AD, Pitt JI, Fleet GH. Growth of fungi and mycotoxin production on cheese under modified atmospheres. Int J Food Microbiol 2001; 68: 125-33.

[190] Taniwaki MH, Hocking AD, Pitt JI, Fleet GH. Growth and mycotoxin production by food spoilage fungi under high carbon dioxide and low oxygen atmospheres. Int J Food Microbiol 2009; 132: 100-8.

[191] González de Llano D, Ramos M, Rodriguez A, Montilla A, Juarez M. Microbiological and physicochemical characteristics of Gamonedo blue cheese during ripening. Int Dairy J 1992; 2: 121-35.

[192] Gripon JC. In: Fox PF, Ed. Cheese: Chemistry, physics and microbiology London: Chapman & Hall. 1993; Vol. 2. pp. 111-136.

[193] Gobbetti M, Burzigotti R, Smacchi E, Corsetti A, de Angelis M. Microbiology and biochemistry of Gorgonzola cheese during ripening. Int Dairy J 1997; 7: 519-29.

[194] Johnson ME. In: Marth EH, Steele JL, Eds. Applied dairy microbiology. 2nd ed. New York, Marcel Dekker, Inc. 2001; pp. 345-384.

[195] Gehrig RF, Knight SG. Formation of ketones from fatty acids by spores of *Penicillium roqueforti*. Nature 1958; 182: 1237.

[196] Kinderlerer JL, Matthias HE, Finner P. Effect of medium-chain fatty acids in mould ripened cheese on the growth of *Listeria monocytogenes*. J Dairy Res 1996; 63: 593-606.

[197] Fox PF, Guinee TP, Cogan TM, McSweeney PLH. Fundamentals of Cheese Science. Gaithersburg, MD: Aspen Publishers Inc. 2000.

[198] Prieto B, Franco I, Fresno JM, Bernardo A, Carballo J. Picón Bejes-Tresviso blue cheese: an overall biochemical survey throughout the ripening process. Int Dairy J 2000; 10: 159-67.

[199] van den Tempel T, Nielsen MS. Effects of atmospheric conditions, NaCl and pH on growth and interactions between moulds and yeasts related to blue cheese production. Int J Food Microbiol 2000; 57: 193-9.

[200] Flórez AB, Mayo B. Microbial diversity and succession during the manufacture and ripening of traditional, Spanish, blue-veined Cabrales cheese, as determined by PCR-DGGE. Int J Food Microbiol 2006; 110: 165-71.

[201] Fox PF, Law J. Enzymology of cheese ripening. Food Biotechnol 1991; 5: 239-62.

[202] Irlinger F, Mounier J. Microbial interactions in cheese: implications for cheese quality and safety. Curr Opin Biotechnol 2009; 20: 142-8.

[203] Beresford TP, Fitzsimons NA, Brennan NL, Cogan TM. Recent advances in cheese microbiology. Int Dairy J 2001; 11: 259-74.

[204] Addis E, Fleet GH, Cox JM, Kolak D, Leung T. The growth, properties and interactions of yeasts and bacteria associated with the maturation of Camembert and blue-veined cheeses. Int J Food Microbiol 2001; 69: 25-36.

[205] Ercolini D, Hill PJ, Dodd CE. Bacterial community structure and location in Stilton cheese. Appl Environ Microbiol 2003; 69: 3540-8.

[206] Flórez AB, Hernández-Barranco AM, Marcos I, Mayo B. Biochemical and microbiological characterization of artisan kid rennet extracts used for Cabrales cheese manufacture. LWT-Food Sci Technol 2006; 39: 605-12.

[207] Roostita R, Fleet GH. The occurrence and growth of yeasts in Camembert and blue-veined cheeses. Int J Food Microbiol 1996; 28: 393-404.

[208] Hansen TK, Jakobsen M. Taxonomical and technological characteristics of *Saccharomyces* spp. associated with blue veined cheese. Int J Food Microbiol 2001; 69: 59-68.

[209] McSweeney PLH, Sousa MJ. Biochemical pathways for the production of flavor compounds in cheeses during ripening: A review. Lait 2000; 80: 293-324.

[210] Larsen MD, Jensen K. The effects of environmental conditions on the lipolytic activity of strains of *Penicillium roqueforti*. Int J Food Microbiol 1999; 46: 159-66.

[211] Taha EE, Knight SG. Physiological and chemical studies on mould proteins. Growth and cell proteins of *Penicillium roqueforti* as influenced by culture conditions. Arch Mikrobiol 1961; 39: 22-36.

[212] van den Tempel T, Nielsen MS. Effects of atmospheric conditions, NaCl and pH on growth and interactions between moulds and yeasts related to blue cheese production. Int J Food Microbiol 2000; 57: 193-9.

[213] van den Tempel T, Gundersen JK, Nielsen MS. The microdistribution of oxygen in Danablu cheese measured by a microsensor during ripening. Int J Food Microbiol 2002; 75: 157-61.

[214] Lawrence RC, Bailey RW. Evidence for the role of the citric acid cycle in the activation of spores of *Penicillium roqueforti*. Biochim Biophys Acta 1970; 208: 77-86.

[215] Godinho M, Fox PF. Effect of NaCl on the germination and growth of *Penicillium roqueforti*. Milchwissenschaft 1981; 36: 205-8.

[216] Fernández-Salguero J, Alcalá M, Marcos A, Esteban MA. Measurement and calculation of water activity in Blue cheese. J Dairy Res 1986; 53: 639-64.

[217] Gock MA, Hocking AD, Pitt JI, Poulos PG. Influence of temperature, water activity and pH on growth of some xerophilic fungi. Int J Food Microbiol 2003; 81: 11-9.

[218] Li Y, Wadsö L, Larsson L. Impact of temperature on growth and metabolic efficiency of *Penicillium roqueforti*-correlations between produced heat, ergosterol content and biomass. J Appl Microbiol 2009; 106: 1494-501.

[219] Fox PF, McSweeney PLH. Proteolysis in cheese during ripening. Food Rev Int 1996; 12: 457-509.

[220] Sousa MJ, Ardö Y, McSweeney PLH. Advances in the study of proteolysis during cheese ripening. Int Dairy J 2001; 11: 327-45.

[221] Grappin R, Rank TC, Olson NF. Primary proteolysis of cheese proteins during ripening. A review. J Dairy Sci 1985; 68: 531-40.

[222] Coker CJ, Crawford RA, Johnston KA, Singh H, Creamer LK. Towards the classification of cheese variety and maturity on the basis of statistical analysis of proteolysis data: a review. Int Dairy J 2005; 15: 631-43.

[223] Larsen MD, Kristiansen KR, Hansen TK. Characterization of the proteolytic activity of starter cultures of *Penicillium roqueforti* for production of blue veined cheeses. Int J Food Microbiol 1998; 43: 215-21.

[224] Farahat SM, Rabie AM, Farag AA. Evaluation of the proteolytic and lipolytic activity of different *Penicillium roqueforti* strains. Food Chem 1990; 36: 169-180.

[225] Gente S, Billon-Grand G, Poussereau N, Févre M. Ambient alkaline pH prevents maturation but not synthesis of ASPA, the aspartyl protease from *Penicillium roqueforti*. Curr Genet 2001; 38: 323-8.

[226] Toelstede S, Hofmann T. Kokumi-active glutamyl peptides in cheeses and their biogeneration by *Penicillium roqueforti*. J Agric Food Chem 2009; 57: 3738-48.

[227] Sablé S, Cottenceau G. Current knowledge of soft cheeses flavor and related compounds. J Agric Food Chem 1999; 47: 4825-36.

[228] Collins YF, McSweeney PLH, Wilkinson MG. Lipolysis and free fatty acid catabolism in cheese: a review of current knowledge. Int Dairy J 2003; 13: 841-66.

[229] Lamberet G, Menassa A. Purification and properties of an acid lipase from *Penicillium roqueforti*. J Dairy Res 1983; 50: 459-68.

[230] Mase T, Matsumiya Y, Matsuura A. Purification and characterization of *Penicillium roqueforti* IAM 7268 lipase. Biosci Biotechnol Biochem 1995; 59: 329-30.

[231] Lamberet G, Menassa A. Détermination et niveau des activités lipolytiques dans les fromages á pate persillée. Le Lait 1983; 63: 333-44.

[232] King RD, Clegg GH. The metabolism of fatty acids, methyl ketones and secondary alcohols by *Penicillium roqueforti* in blue cheese slurries. J Sci Food Agric 1979; 30: 197-202.

[233] Madkor S, Fox PF, Shalabi SI, Metwalli NH. Studies on the ripening of stilton cheese: Lipolysis. Food Chem 1987; 25: 93-109.

[234] Contarini G, Toppino PM. Lipolysis in Gorgonzola cheese during ripening. Int Dairy J, 1995, (2): 141-155.

[235] Cox AD. Farnesyltransferase inhibitors: potential role in the treatment of cancer. Drugs 2001; 61: 723-32.

[236] Shiomi K, Uchida R, Inokoshi J, Tanaka H, Iwai Y, Omura S. Andrastins A~C, new protein farnesyltransferase inhibitors, produced by *Penicillium* sp. FO-3929. Tetrahedron Lett 1996; 37: 1265-8.

[237] Uchida R, Shiomi K, Inokoshi J, *et al.* Andrastins A-C, new protein farnesyltransferase inhibitors produced by *Penicillium* sp. FO-3929. II. Structure elucidation and biosynthesis. J Antibiot (Tokyo) 1996; 49: 418-24.

[238] Omura S, Inokoshi J, Uchida R, *et al.* Andrastins A-C, new protein farnesyltransferase inhibitors produced by *Penicillium* sp. FO-3929. I. Producing strain, fermentation, isolation, and biological activities. J Antibiot (Tokyo) 1996; 49: 414-7.

[239] Nielsen KF, Dalsgaard PW, Smedsgaard J, Larsen TO. Andrastins A-D, *Penicillium roqueforti* metabolites consistently produced in blue-mold-ripened cheese. J Agric Food Chem 2005; 53: 2908-13.

[240] O'Brien M, Nielsen KF, O'Kiely P, Forristal PD, Fuller HT, Frisvad JC. Mycotoxins and other secondary metabolites produced *in vitro* by *Penicillium paneum* Frisvad and *Penicillium roqueforti* Thom isolated from baled grass silage in Ireland. J Agric Food Chem 2006; 54: 9268-76.

[241] Sonjak S, Frisvad JC, Gunde-Cimerman N. Comparison of secondary metabolite production by *Penicillium crustosum* strains, isolated from Arctic and other various ecological niches. FEMS Microbiol Ecol 2005; 53: 51-60.

[242] Overy DP, Larsen TO, Dalsgaard PW, *et al.* Andrastin A and barceloneic acid metabolites, protein farnesyl transferase inhibitors from *Penicillium albocoremium*: chemotaxonomic significance and pathological implications. Mycol Res 2005; 109: 1243-9.

[243] Haluska P, Dy GK, Adjei AA. Farnesyl transferase inhibitors as anticancer agents. Eur J Cancer 2002; 38: 1685-700.

[244] Sun J, Qian Y, Hamilton AD, Sebti SM. Both farnesyltransferase and geranylgeranyltransferase I inhibitors are required for inhibition of oncogenic K-Ras prenylation but each alone is sufficient to suppress human tumor growth in nude mouse xenografts. Oncogene 1998; 16: 1467-73.

[245] Kato K, Cox AD, Hisaka MM, Graham SM, Buss JE, Der CJ. Isoprenoid addition to Ras protein is the critical modification for its membrane association and transforming activity. Proc Natl Acad Sci USA 1992; 89: 6403-7.

CHAPTER 6

Biological Activity and Production of *Monascus* Metabolites

Yii-Lih Lin, Teng-Hsu Wang, Min-Hsiung Lee and Nan-Wei Su*

Department of Agricultural Chemistry, National Taiwan University, Taiwan

Abstract: *Monascus* fermented products are conventional food colorant and seasoning in many Asian countries. In recent years, they were found to ameliorate several civilization diseases including hyperlipidemia, hypercholesterolemia, and hypertension. Therefore, *Monascus* related products are popular on the market and are vastly consumed as a dietary supplement. Monacolin K is one of its bioactive metabolites and has been approved as a clinical prescription named Lovastatin. *Monascus* can be a potential source of various bioactive compounds. In this chapter, we summarize these bio-functional metabolites and also review the process and factors controlling the production of *Monascus* related products.

INTRODUCTION

Toxicology and side effects of folk medicine products remain a considerable risk as they are not sufficiently tested. On the other hand, *Monascus* related goods, as a traditional source of food and medicinal products, have been used for thousands of years by a vast majority of population, and are therefore considered "generally recognized as safe."

In Asian countries, such as China, Japan, Taiwan, Korea, Thailand, Philippines, and Indonesia, *Monascus* fermented rice is a traditional food colorant and a preservative for fish and meat. It is also used as a flavoring ingredient for a variety of Chinese dishes [1]. For example, roasted pork and duck are frequently covered with a layer of *Monascus* fermented rice, rendering the unique complex flavor, aromatic fragrance, and vivid red color. *Monascus* is also used in the starter culture for the brewing of red colored liquors. *Monascus* fermented rice has several synonyms, including Hung-Chu, Hong-Qu, Ang-kak, Ankak rice, Red Yeast rice, Red Mold rice, and Beni-Koji. The rice is also described as a mild folk medicine in an ancient Chinese pharmacopedia, Ben-Cao-Gang-Mu, which was composed by Shi-Zhen Li (1518-1593 A.D.).

In biological taxonomy, the genus *Monascus* belongs to the family Monascaceae, the order Eurotiales, the class Ascomycetes, the phylum Ascomycota, and the kingdom Fungi [2]. *Monascus* was identified and named by Van Tieghem in 18th century [3]. After ten years, *Monascus purpureus* was identified on the brightly colored rice grains [4]. Currently, 67 *Monascus* strains are deposited in American Type Culture Collection, although based on Hawksworth and Pit's work, most strains belong to only three species: *M. pilosus; M. purpureus; M. ruber* [5].

M. purpureus is the major species used as seasonings or dietary supplements. They are also known as a synonym of *M. anka* because it was isolated from the Ankak rice in Taiwan in 1930 [6, 7].

BIOACTIVITY OF *MONASCUS* CRUDE EXTRACT

The crude extract of *Monascus* fermented product (MFP) demonstrates significant bioactivity without further isolation process. It has been reported as a bio-functional dietary supplement to reduce the level of plasma glucose, cholesterol, and triacylglyceride. Rats with oral administration of MFP demonstrate release of acetylcholine, which in turn promotes insulin release, and thus reduces the plasma glucose [8]. For streptozotocin-induced diabetic rats, oral administration of MFP decreases their plasma glucose. For normal rats subjected to intravenous glucose injection, oral administration of MFP also attenuates the elevation of plasma glucose [9]. In a clinical study, 79 patients with hyperlipidemia were randomly and double-blinded grouped to receive MFP or placebo daily. The patients, after eight weeks of MFP administration, showed reduced levels of low-density lipoprotein cholesterol, total cholesterol, triglycerides and apolipoprotein B [10]. Interestingly, feeding hens with MFP not only reduced

*Address correspondence to Nan-Wei Su: Department of Agricultural Chemistry, National Taiwan University, Taiwan; E-mail: snw@ntu.edu.tw

Ana Lúcia Leitão (Ed)

their levels of cholesterol, triglycerides, and LDL in serum, but also decreased the cholesterol content in egg yolk [11, 12]. This finding not only confirms the bioactivity of MFP, but also provides a healthier source of meat and egg products for consumers who demand to control their cholesterol uptake.

The red *Monascus* fermented rice also exhibit potent activity to ameliorate Alzheimer's disease in animal models. When the ethanol extract of *Monascus* fermented rice were fed to rats with Alzheimer's disease, they demonstrated improved memory and learning abilities in both the water maze task and in another experiment to avoid electric shock [13]. The underlying mechanism is complex, and could be the result of different ingredients. One report indicated that the extract with various anti-oxidants, majorly monacolin K, exhibited anti-oxidation and anti-inflammatory effects to cells, and thus repressed the neurotoxicity from amyloid beta peptide [14]. Another research indicated that in animal model, the neuron protective activity was provided by down regulation of amyloid beta 40 formation and deposition, suppressing the expression of cholesterol-raided β-secretase activity and apolipoprotein E expression, and also promoting the secretion of neuron-protective protein, APP α-fragment (sAAPα) in hippocampus.

Although the composition of MFP is complicated, its bio-function has been studied in model systems preliminarily, shedding light on its biological mechanism. For example, MFP extract significantly decreases the gene expressions and enzyme activity related to adipocyte differentiation. Briefly, glycerol-3-phosphate dehydrogenase activity and lipid accumulation are decreased in mammalian cells, and the mRNA levels of CCAAT/enhancer-binding protein (C/EBP) and peroxisome proliferator-activated receptor (PPAR) are decreased. These results suggest an inhibitory function of MFP to the susceptibility of adipocyte differentiation [15]. Moreover, MFP extracts reduced the holocysteine-stimulated endothelial adhesiveness as well as its downstream signaling transduction, including intracellular reactive oxygen species (ROS), the nuclear transcription factor NF-κB expression and vascular cell adhesion molecule-1 (VCAM-1) expression in human aoric endothelial cells. These results imply a potential application of MFP in clinical atherosclerosis disease [16].

On the way to dissect the complexity the MFP was repetitively fractionated and isolated. Follow its bio-active fraction and eventually identify the active single compound. γ-aminobutyric acid (GABA) (Fig. **3d**) was found in a systematic fractionation and isolation process, searching for the bio-active compound in MFP [17]. GABA, one of the major inhibitory neurotransmitters in the central nervous system, was reported to reduce blood pressure in animal model and human [18-21]. Furthermore, GABA-rich foods have also been reported to effectively reduce blood pressure in spontaneous hypertensive rats [22, 23]

BIOACTIVITY OF PIGMENTS IN *MONASCUS*

Most of the MFP carries deep red color, indicating its rich content of colorants. There are eight major pigments (Figs. **1a, 1b, 1c, 1d, 1e, 1f, 1g, 1h**). The red colorants named rubropunctamine (Fig. **1e**) and monascorubramine (Fig. **1f**) are the most abundant. The orange colorants are rubropunctatin (Fig. **1a**) and monascorubrin (Fig. **1b**). The yellow colorants are monascin (Fig. **1c**) and ankaflavin (Fig. **1d**). Moreover, a yellowish colorant named Xanthomonasin A (Fig. **1g**) in the mutant of *Monascus anka* was identified [24]. Due to the improvement of photodiode array and fluorescence detector, more colorant compounds have been identified [25].

Yasukawa *et al.* used 12-O-tetradecanoyl-phorbol-13-acetate (TPA) as an inflammatory agent to promote carcinogenesis on mouse ear. They found that pigment mixtures from *Monascus anka* were able to suppress the tumor formation process [26]. Among the pigments, monascorubrin is the most effective one and its function is assumed through its anti-inflammatory activity [27]. In another report regarding skin cancer on mouse, oral administration of monascin inhibits the carcinogenesis of skin cancer initiated by peroxynitrite or ultraviolet light, followed by TPA promotion [28]. Moreover, ankaflavin, shows selective cytotoxicity to cancer cell lines by apoptosis-related mechanism, and shows relatively low toxicity to normal fibroblasts. Its structural analog monascin, however, shows no cytotoxicity to all cell lines tested [29]. The orange pigments, rubropunctatin and monascorubrin, shows antibiotic activity against bacteria, yeast, and filamentous fungi [30]. Rubropunctatin and monascorubrin could inhibit the growth of *Bacillus subtilis* and *Candida pseudotropicalis*.

The stability of *Monascus* pigments are significantly affected by light exposure, temperature, and pH value [31]. In a pigment stability test, in which the pigments were prepared by methanol:chloroform (1:1) extraction of freeze-dried culture broth of *M. ruber* van Tieghem, after light exposure for 50 days, a decay of 20% was observed. Moreover, a pigment decay of 30% was observed after a treatment at 100 °C during 8 hours. The red pigment was found more stable in neutral (pH 7) or alkaline (pH 9.5) than in acidic condition (pH 3). The authors further investigated the applicability of these pigments to replace conventional colorants in meat products (such as nitrite salts or cochineal). Based on the results of sensory evaluation, the *Monascus* colorants incorporated meat products exhibited stable color (95% stability after 3 months at 4 °C under vacuum) and enhanced flavor and texture. Therefore, the authors concluded that *Monascus* pigments were superior preservatives than nitrite salts and could be used as a suitable substitute for food additives in meat products.

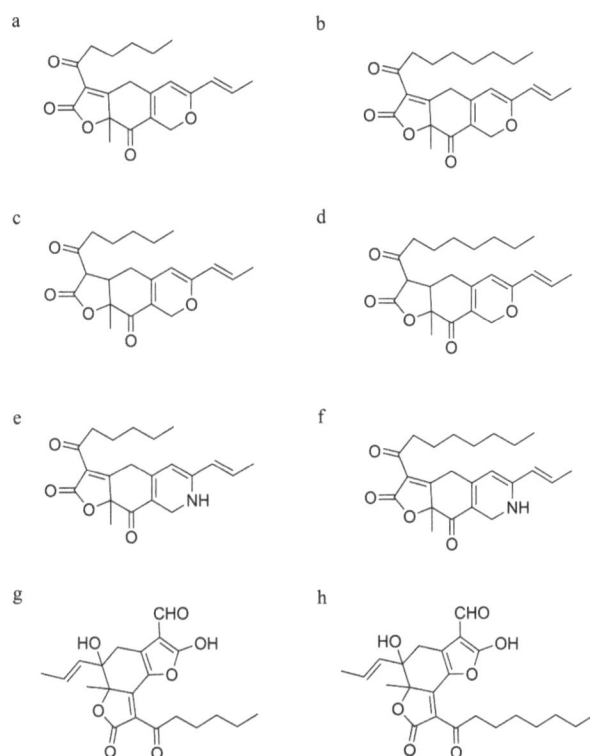

Figure 1: Chemical structures of pigments from *Monascus*: rubropunctatin (a), monascorubrin (b), monascin (c), ankaflavin (d), rubropunctamine (e), monascorubramine (f), xanthomonasin A (g) and xanthomonasin B (h).

BIOACTIVITY RELATED TO CHOLESTEROL AND LIPID METABOLISM

Among all the bioactive compounds in MFP, monacolins (Fig. **2**) are the most well-known due to their pharmacological effects to control hyperlipidemia [30, 31]. Among monacolins, monacolin K is an efficacious compound to lower cholesterol level. It is also named as lovastatin, mevinolin, and mevacor [32]. Current therapeutic strategy to reduce the cholesterol levels is to block the activity of 3-hydroxy-3-methylglutaryl coenzyme A (HMG-CoA) reductase, which is the rate determining enzyme of cholesterol synthesis pathway. Compactin (ML-236B) was the first discovered inhibitor to HMG-CoA reductase [36, 37]. Afterward, monacolin K (mevinolin) was found as a structurally similar but more effective compound than compactin. It was purified from the metabolites of *Monascus ruber* and *Aspergillus terreus* [32, 33, 38], and was further commercialized by Merck & Co., Inc. in the name of Lovastatin. Structural analogs including Monacolin J, L, and M, were also found capable to reduce cholesterol synthesis [39, 40].

BIOACTIVITY RELATED TO ANTI-OXIDATION

Dihydromonascolins are structural analogs to monacolins (Fig. **3a-b**). Dihydromonacolin-MV (Fig. **3b**), which is derived from the methanolic extract of *Monascus purpureus*, exhibits strong anti-oxidative activity in the 2,2-diphenyl-1-picrylhydrazyl (DPPH) radical scavenging assay and is able to inhibit lipid peroxidation in a liposome model [41].

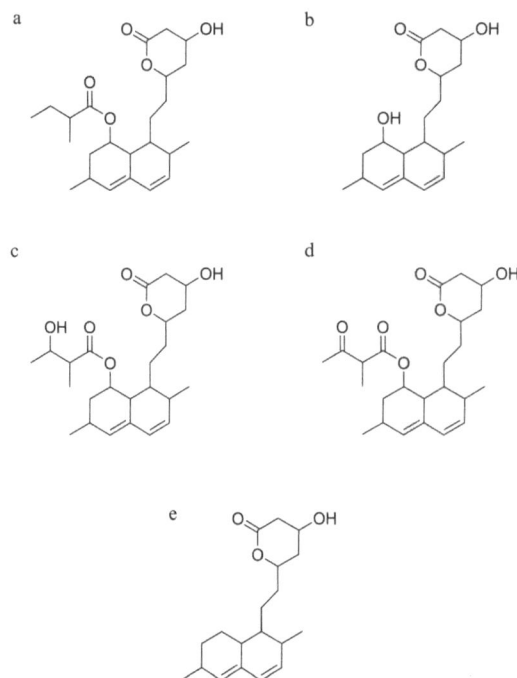

Figure 2: Chemical structures of monacolins: monacolin K (a), monacolin J (b), monacolin M (c), monacolin X (d), monacolin L (e).

Dimerumic acid (Fig. **3e**) also shows an *in vitro* anti-oxidative activity in the DPPH assay, and was identified as the major constituent responsible for the anti-oxidative and hepatoprotective activity in *Monascus* extract in the liver injured mice induced by carbon tetrachloride [42, 43]. Dimerumic acid is further found to inhibit the NADPH- and iron(II)-dependent lipid peroxidation of rat liver microsomes. The antioxidative property is contributed by the electron donation of the hydroxamic acid group to the oxidants [44]. Furthermore, dimerumic acid is found to inhibit salicylic acid and tert-butylhydroperoxide induced oxidation and cytotoxicity, in a study using rat liver microsomes and isolated hepatocytes [45].

TOXIC COMPOUNDS IN *MONASCUS*

Citrinin (Fig. **3c**) was initially named as monascidin A and was regarded as an antibacterial component in the crude extract of *Monascus*. Monascidin A was then confirmed to be the same compound as citrinin [46]. Citrinin is the hepatotoxic and nephrotoxic compound in MFP. It adversely affects the function and ultrastructure of kidney in canine [47]. Citrinin also have negative effects on serum glucose [48]. Although the detailed molecular mechanism of the toxicity of citrinin requires more investigations, it has been demonstrated that citrinin mainly affects mitochondria in cells. Citrinin permeates into the mitochondria, alters Ca^{2+} homeostasis [49], and interferes with the electron transport system [50]. Citrinin is not a mutagen itself; however, if it is transformed by hepatocytes, it becomes mutagenic to fibroblast cells. The citrinin content dominates the mutagenicity of *Monascus* fermentation products in a dosage dependent manner. Only the samples with higher citrinin content show positive mutagenic response in the *Salmonella*-hepatocyte assay [51]. Citrinin has also been reported as a teratogenic agent in chicken embryos [52].

Although MFP can be easily purchased in the market without any restrictions, its adverse compound citrinin is still a concern. Therefore, efforts have been made to decrease the content of citrinin. Although citrinin is synthesized through polyketide pathway, through which many secondary metabolites are synthesized, especially pigments, the synthesis of pigments and citrinin are not necessarily correlated [53]. Some of the *Monascus* strains produce pigments without citrinin [54]. Yet the production of monacolin K without the existence of citrinin is not possible. Until now, we can only screen for the most suitable strain and optimize the parameters in production, to reduce the citrinin content in order to pass the statutory threshold. Currently, Japanese government have issued legislated restrictions, defining the statutory limit of 0.2 µg/g (200 ppb) of the *Monascus* pigments as used in food additives [55]. Taiwan government also declared a new legislated regulation on citrinin content in food products in December,

2009. The citrinin content in the *Monascus* pigments should be lower than 200 ppb. In the red *Monascus*-fermented rice, it should be less than 5 ppm, and should not exceeds 2 ppm in the final food products.

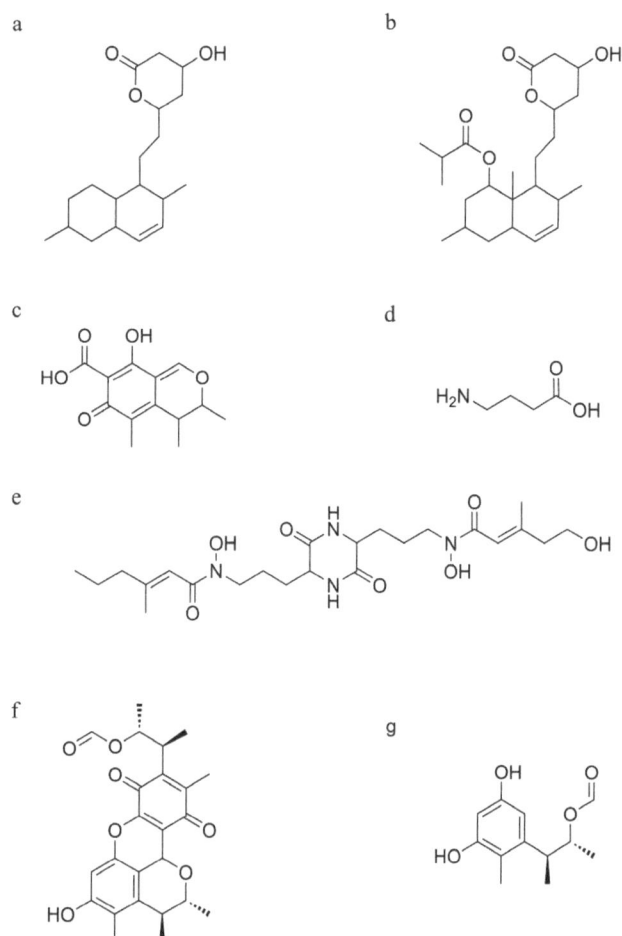

Figure 3: Chemical structures of other metabolites from *Monascus*: dihydromonacolin-L (a), dihydromonacolin-MV (b), citrinin (c), γ-aminobutyric acid (d), dimerumic acid (e), citrinin H1 (f), citrinin H2 (g).

Most molecules decompose in high temperature environment, including citrinin, which decomposes and loses its cytotoxicity to mammalian cells after treated with 175 °C dry air. Humidification lowers the temperature required to deactivate the cytotoxicity of citrinin. In a moisten environment (200 g citrinin/150 g H_2O) the deactivation temperature is reduced to 160-175 °C [56]. Citrinin H2 (Fig. **3g**), which is less toxic than citrinin, is considered the major product of citrinin decomposition [57]. Although the thermal process facilitates citrinin decomposition, a compound, citrinin H1 (Fig. **3f**), forms with ten folds of toxicity (on a weight basis) [58].

SOLID STATE FERMENTATION

In order to acquire high consistency in quality, especially in the case of large scale production, some manufactures have implemented laboratory sterilization processes and automated facilities. Most of them use the solid state fermentation procedure and thus the qualities of products are determined by the following parameters: substrates (predominantly non-glutinous rice kernel), selected *Monascus* strains, temperature, moisture content of fermentation mixture along the process, and control of contamination factors [59].

Until monacolins and GABA were reported, good production quality of MFP was long been correlated with higher pigment accumulation in the fermentation mixture. Juzlova *et al.* summarized several articles in a review which addressed solid state cultivation of *Monascus* sp. in laboratorial scale for pigment production [60].

Among rice substrates used in producing MFP, non-glutinous soft rice contains more water content and more available starch to molds comparing to glutinous rice [7, 61]. Adlay is reported an alternative fermentation substrate. It is a common crop in Eastern Asian and also a traditional Chinese herb. Comparing to the rice substrate, *Monascus* culture on adlay produces less citrinin (0.26-14.64 ppm) [61].

Water content is the first factor to control during MFP production. It is reported that the optimal substrate humidity should be approximately 40-50% at the beginning of the process and maintained by temporarily moistening the substrates in order to favor fungal growth [62]. Recent studies suggested that lower initial moisture content (25-30%) helps to keep a low glucoamylase activity, and thus is beneficial to the pigment yields [63, 64]. However, Chen *et al.* concluded that cultivation of a mutant strain *M. pilosus* M12-69 yielded the optimized monacolin K/citrinin ratio when the water content is between 55% and 75% [65].

Sufficient aeration is also important to pigment production. Pigment formation was dramatically blocked when excess of carbon dioxide accumulated in incubator [64]. In the Chinese ancient process, sufficient aeration is achieved by stirring the fermentation mixture on bamboo trays every two hours to separate grains from agglomerates. In the laboratory scale, the separation is substantially carried out by shaking the substrates in flasks or plastic bags. Recently, a commercial koji maker with a rotary perforated bed with a diameter of five meter was adopted for MFP mass production [66]. The machine provides up-flow aeration and a plowing mixer to facilitate heat dissipation and prevent agglomeration.

Optimization of the composition of ingredients, in other words, is to increase the beneficial ingredients and decrease the toxic components. A genetically modified strain of *Monascus* with higher monacolin K productivity and lower citrinin content, and obtained by random mutation has been described elsewhere [67]. In the strain *Monascus purpureus* NTU 601, addition of 0.5% of ethanol as carbon source tripled the content of monacolin K, increased the GABA production to seven fold, and reduced the citrinin content [68]. Moreover, substrates for preferential production of certain metabolites have also been studied. For example, *Dioscorea batatas* is reported as an enhancer substrate for *Monascus* species to the production of monacolin K and monascin (Figs. **2c** and **1a**) [69].

LIQUID/SUBMERGED CULTIVATION

Liquid/submerged cultivation as an alternative methodology to *Monascus* cultivation provides a stringent control of cultivation parameters. Various culturing parameters including water supplement, temperature, nitrogen source, medium components, and pH value have been investigated [70, 71]. Lin was the first to study liquid culture conditions on the pigment production by *Monascus* [72, 73]. The cultural conditions for maximum pigmentation were 5% rice powder (with 3.5% starch content) as carbon source; 0.5% of sodium nitrate or potassium nitrate as nitrogen source; initial pH of 6.0; and the temperature of 32 °C. Su and Huang reported that polished rice powder gave higher pigment production and *Monascus* dry weight than rice starch, indicating minor compositions in rice kernel benefit fungal growth and pigmentation. Comparing to rice powder, corn powder used as substrate gave higher fungal dry weight but poor pigment production. They also suggested that addition of 1-2% alcohol during incubation has favorable effect on pigment production [62]. Carels and Shepherd investigated the effect of different nitrogen sources on final culture acidity and the pigment content [74]. During cultivation, the final medium pH was around 6.5 when yeast extract or nitrate was used as nitrogen source. However, when ammonium or ammonium nitrate was used, the pH decreased to around 2.5 and the pigments became orange. Chen and Johns further examined the effect of pH and nitrogen source on individual pigment productions in a pH-controlled fermentor [75]. Lower pH value (pH 4.0) promotes the fungal growth and favors the synthesis of ankaflavin (Fig. **1d**); however, the production of other pigments is relatively non-susceptible to pH value.

Oxygen concentration in submerged cultivation also affects the biosynthesis of *Monascus* metabolites [76]. By adjusting dissolved oxygen concentration and agitation speed, optimized red pigment/citrinin production ratio were obtained [77]. In oxygen-limiting incubation, production of pigments (Fig. **1**) and citrinin (Fig. **3c**) are growth-related and are both biosynthesized as primary metabolites. In oxygen-excess condition, however, citrinin is produced as a secondary metabolite, which is mostly produced during the stationary phase. In contrast, the pigments decrease dramatically during incubation. Pigments formation is partially inhibited by metabolites produced in aerobic environments such as L-maltose, succinate, and dicarboxylic acid; however, the formation of citrinin is not affected.

Other components in cultural medium were also reported to affect pigment productions. For instance, addition of leucine to the culture medium interferes with the production of red pigment [78]. The absence of potassium phosphate in the medium depresses the red pigment production in the culture of *Monascus pilosus* [79].

Interestingly, spectrum of light also affects the composition of *Monascus* secondary metabolites [80]. Miyake *et al.* incubated the *Monascus* in shaking flasks, either in the dark or exposed it to the red light (635 nm) or blue light (470 nm). The red light promotes the production of red pigments and citrinin, and the blue light promotes the production of GABA. Furthermore, the spectrum of light also affects mycelium development and spore formation of *Monascus*.

TECHNICAL PERSPECTIVE ON THE PRODUCTION PROCESS

Along with the progress of current biotechnology, many of the state-of-the-art methodology are converged to explore the genomic information as well as the mechanism of secondary metabolite formation. Systems biology approaches are recently introduced to elucidate the complex pathways of biosynthesis. For instance, proteomic analysis with MALDI-TOF and MS/MS has revealed that the phosphate limitation condition up-regulates the expression of aldehyde dehydrogenase and the glycolytic enzymes [79]. Moreover, the genomic database of *Monascus* sp. has been established and is currently maintained in Bioresource Collection and Research Center, Taiwan, although this database is not currently opened for public. In 2008, a putative gene cluster for the biosynthesis of monacolin K was identified. Nine genes designated as *mok*A to *mok*I, which share over 54% similarity with lovastatin biosynthetic gene cluster in *Aspergillus terreus* were reported [81].

An alternative way to acquire beneficial or coloring ingredient without coproduction of citrinin is to screen for an alternative microorganism producer. Computer screening was conducted to search for *Monascus*-like pigments from ascomycetous filamentous fungi belonging to *Penicillium* sp. The pioneering approach opens a door to biosynthetically acquire *Monascus*-like ingredients without citrinin [82]. Although genetic manipulation to *Monascus ruber* is not as well-established as typical model organisms, gene delivery has been achieved through the mediation of *Agrobactrium tumefaciens* [83].

CONCLUSION

MFP has been widely used in Asia as a natural food colorant and additive. It is already commercialized as a dietary supplement to ameliorate hypertention, hypercholesterolemia and hyperlipidemia. Among its various biofunctional constituents, monacolin K is practically used as a medicine to reduce the blood cholesterol concentration. The production of *Monascus* related products could be improved by selection of the strains, modification of incubation conditions, and genetic engineering to reduce the content of citrinin as well as to increase the production of monacolin K and GABA. Although MFP is well accepted as a dietary supplementary, its complexity in constituents and its citrinin content are still concerned.

REFERENCES

[1]   Wang T, Lin T, Steve LT. *Monascus* Rice Products. In: Taylor S, Ed. Advances in Food and Nutrition Research. London: Academic Press 2007; pp. 123-59.

[2]   Young EM. Physiological studies in relation to the taxonomy of *Monascus* spp. In: Juday C, Ed. Transactions of the Wisconsin Academy of Sciences, Arts and Letters. Madison, Wis., 1930; pp. 227-plate 4ff.

[3]   Tieghem MV. *Monascus* genre nouvear de l'ondre des Ascomycetes. Acta Bot Gallica 1884; 31: 226-31.

[4]   Went FAFC. *Monascus purpureus*, le champignon de l'Ang-Quac, une nouvelle Thélébolée. Ann Sci Natl Bot 1895; 1: 1-18.

[5]   Hawksworth DL, Pit JI. A new taxonomy for *Monascus* species based on cultural and microscopical characters. Aust J Bot 1983; 31(1): 51-61.

[6]   Su YC, Chen WL, Fang HY, Wong HC, Wang WH. Mycological study of *Monascus anka* (in Chinese). J Chin Agric Chem Soc 1970; 8(1~2): 46-54.

[7]   Steinkraus KH, Ed. Handbook of indigenous fermented foods. New York M: Dekker 1983.

[8]   Chen CC, Liu IM. Release of acetylcholine by Hon-Chi to raise insulin secretion in Wistar rats. Neurosci Lett 2006; 404(1-2): 117-21.

[9]   Chang JC, Wu MC, Liu IM, Cheng JT. Plasma glucose-lowering action of Hon-Chi in streptozotocin-induced diabetic rats. Horm Metab Res 2006; 38(2): 76-81.

[10]   Lin CC, Li TC, Lai MM. Efficacy and safety of *Monascus purpureus* Went rice in subjects with hyperlipidemia. Eur J Endocrinol 2005; 153(5): 679-86.

[11]   Wang JJ, Pan TM, Shieh MJ, Hsu CC. Effect of red mold rice supplements on serum and meat cholesterol levels of broilers chicken. Appl Microbiol Biotechnol 2006; 71(6): 812-8.

[12]   Wang JJ, Pan TM. Effect of red mold rice supplements on serum and egg yolk cholesterol levels of laying hens. J Agric Food Chem 2003; 51(16): 4824-9.

[13]   Lee CL, Kuo TF, Wang JJ, Pan TM. Red mold rice ameliorates impairment of memory and learning ability in intracerebroventricular Amyloid β-infused rat by repressing amyloid β accumulation. J Neurosci Res 2007; 85(14): 3171-82.

[14]   Lee CL, Wang JJ, Pan TM. Red mold rice extract represses amyloid beta peptide-induced neurotoxicicity *via* potent synergism of anti-inflammatory and antioxidative effect. Appl Microbiol Biotechnol 2008; 79(5): 829-41.

[15]   Jeon T, Hwang SG, Hirai S, *et al.* Red yeast rice extracts suppress adipogenesis by down-regulating adipogenic transcription factors and gene expression in 3T3-L1 cells. Life Sci 2004; 75(26): 3195-203.

[16]   Lin CP, Chen YH, Chen JW, *et al.* Cholestin (*Monascus purpureus* rice) inhibits homocysteine-induced reactive oxygen species generation, nuclear factor-kappa B activation, and vascular cell adhesion molecule-1 expression in human aortic endothelial cells. J Biomed Sci 2008; 15(2): 183-96.

[17]   Kohama Y, Matsumoto S, Mimura T, Tanabe N, Inada A, Nakanishi T. Isolation and identification of hypotensive principles in red-mold rice. Chem Pharm Bull 1987; 35(6): 2484-9.

[18]   Watanabe M, Maemura K, Oki K, Shiraishi N, Shibayama Y, Katsu K. Gamma-aminobutyric acid (GABA) and cell proliferation: focus on cancer cells. Histol Histopathol 2006; 21(10): 1135-41.

[19]   Kerr DIB, Ong J. GABA$_B$ receptors. Pharmacol Ther 1995; 67(2): 187-246.

[20]   Blein S, Hawrot E, Barlow P. The metabotropic GABA receptor: molecular insights and their functional consequences. Cell Mol Life Sci 2000; 57(4): 635-50.

[21]   Hayakawa K, Kimura M, Kamata K. Mechanism underlying γ-aminobutyric acid-induced antihypertensive effect in spontaneously hypertensive rats. Eur J Pharmacol 2002; 438(1-2): 107-13.

[22]   Aoki H, Furuya Y, Endo Y, Fujimoto K. Effect of γ-aminobutyric acid-enriched tempeh-like fermented soybean (GABA-Tempeh) on the blood pressure of spontaneously hypertensive rats. Biosci Biotechnol Biochem 2003; 67(8): 1806-8.

[23]   Hayakawa K, Kimura M, Matsumoto K, Sansawa H, Yamori Y. Effect of a γ-aminobutyric acid-enriched dairy product on the blood pressure of spontaneously hypertensive and normotensive Wistar-Kyoto rats. Br J Nutr 2004; 92(3): 411-7.

[24]   Martinkova L, Patakova-Juzlova P, Kren V, *et al.* Biological activities of oligoketide pigments of *Monascus purpureus*. Food Addit Contam 1999; 16(1): 15-24.

[25]   Zheng YQ, Xin YW, Guo YH. Study on the fingerprint profile of *Monascus* products with HPLC-FD, PAD and MS. Food Chem 2008; 113(2): 705-11.

[26]   Yasukawa K, Akihisa T, Oinuma H, *et al.* Inhibitory effect of taraxastane-type triterpenes on tumor promotion by 12-O-tetradecanoylphorbol-13-acetate in two-stage carcinogenesis in mouse skin. Oncology 1996; 53(4): 341-4.

[27]   Yasukawa K, Takahashi M, Natori S, *et al.* Azaphilones inhibit tumor promotion by 12-O-Tetradecanoylphorbol-13-acetate in 2-stage carcinogenesis in mice. Oncology 1994; 51(1): 108-12.

[28]   Akihisa T, Tokuda H, Ukiya M, *et al.* Anti-tumor-initiating effects of monascin, an azaphilonoid pigment from the extract of *Monascus pilosus* fermented rice (red-mold rice). Chem Biodiver 2005; 2(10): 1305-9.

[29]   Su NW, Lin YL, Lee MH, Ho CY. Ankaflavin from *Monascus*-fermented red rice exhibits selective cytotoxic effect and induces cell death on Hep G2 cells. J Agric Food Chem 2005; 53(6): 1949-54.

[30]   Martinkova L, Juzlova P, Vesely D. Biological-activity of polyketide pigments produced by the fungus *Monascus*. J Appl Bacteriol 1995; 79(6): 609-16.

[31]   Fabre CE, Santerre AL, Loret MO, *et al.* Production and food applications of the red pigments of *Monascus ruber*. J Food Sci 1993; 58(5): 1099-110.

[32]   Endo A. Monacolin-K, a new hypocholesterolemic agent produced by a *Monascus* species. J Antibiot 1979; 32(8): 852-4.

[33]   Endo A. Monacolin-K, a new hypocholesterolemic agent that specifically inhibits 3-hydroxy-3-methylglutaryl coenzyme a reductase. J Antibiot 1980; 33(3): 334-6.

[34]   Akihisa T, Tokuda H, Yasukawa K, *et al.* Azaphilones, furanoisophthalides, and amino acids from the extracts of *Monascus pilosus*-fermented rice (red-mold rice) and their chemopreventive effects. J Agric Food Chem 2005; 53(3): 562-5.

[35]   Mabuchi H, Haba T, Tatami R, *et al.* Effects of an inhibitor of 3-hydroxy-3-methylglutaryl coenzyme-a reductase on serum-lipoproteins and ubiquinone-10 levels in patients with familial hypercholesterolemia. N Engl J Med 1981; 305(9): 478-82.

[36] Yamamoto A, Sudo H, Endo A. Therapeutic effects of Ml-236b in primary hypercholesterolemia. Atherosclerosis 1980; 35(3): 259-66.

[37] Alberts AW, Chen J, Kuron G, *et al.* Mevinolin: A highly potent competitive inhibitor of hydroxymethylglutaryl-coenzyme a reductase and a cholesterol-lowering agent. Proc Natl Acad Sci USA 1980; 77(7): 3957-61.

[38] Endo A, Hasumi K, Nakamura T, Kunishima M, Masuda M. Dihydromonacolin-L and Monacolin-X, new metabolites those inhibit cholesterol-biosynthesis. J Antibiot 1985; 38(3): 321-7.

[39] Endo A, Hasumi K, Negishi S. Monacolin-J and Monacolin-L new inhibitors of cholesterol-biosynthesis produced by *Monascus ruber*. J Antibiot 1985; 38(3): 420-2.

[40] Endo A, Hasumi K, Yamada A, Shimoda R, Takeshima H. The synthesis of compactin (Ml-236b) and Monacolin-K in fungi. J Antibiot 1986; 39(11): 1609-10.

[41] Dhale MA, Divakar S, Kumar SU, Vijayalakshmi G. Isolation and characterization of dihydromonacolin-MV from *Monascus purpureus* for antioxidant properties. Appl Microbiol Biotechnol 2007; 73(5): 1197-202.

[42] Aniya Y, Yokomakura T, Yonamine M, *et al.* Screening of antioxidant action of various molds and protection of *Monascus anka* against experimentally induced liver injuries of rats. Gen Pharmacol - the Vasc Syst 1999; 32(2): 225-31.

[43] Aniya Y, Ohtani II, Higa T, *et al.* Dimerumic acid as an antioxidant of the mold, *Monascus anka*. Free Radic Biol Med 2000; 28(6): 999-1004.

[44] Taira J, Miyagi C, Aniya Y. Dimerumic acid as an antioxidant from the mold, *Monascus anka*: the inhibition mechanisms against lipid peroxidation and hemeprotein-mediated oxidation. Biochem Pharmacol 2002; 63(5): 1019-26.

[45] Yamashiro JI, Shiraishi S, Fuwa T, Horie T. Dimerumic acid protected oxidative stress-induced cytotoxicity in isolated rat hepatocytes. Cell Biol Toxicol 2008; 24(4): 283-90.

[46] Blanc PJ, Laussac JP, Lebars J, *et al.* Characterization of Monascidin A from *Monascus* as citrinin. Int J Food Microbiol 1995; 27(2-3): 201-13.

[47] Krejci ME, Bretz NS, Koechel DA. Citrinin produces acute adverse changes in renal function and ultrastructure in pentobarbital-anesthetized dogs without concomitant reductions in [potassium](plasma). Toxicology 1996; 106(1-3): 167-77.

[48] Chagas GM, Oliveira MBM, Campello AP, Kluppel M. Mechanism of citrinin-induced dysfunction of mitochondria.2. Effect on respiration, enzyme-activities, and membrane-potential of liver-mitochondria. Cell Biochem Funct 1992; 10(3): 209-16.

[49] Chagas GM, Oliveira MBM, Campello AP, Kluppel MLW. Mechanism of citrinin-induced dysfunction of mitochondria.4. Effect on Ca2+ transport. Cell Biochem Funct 1995; 13(1): 53-9.

[50] Ribeiro SMR, Chagas GM, Campello AP, Kluppel MLW. Mechanism of citrinin-induced dysfunction of mitochondria.5. effect on the homeostasis of the reactive oxygen species. Cell Biochem Funct 1997; 15(3): 203-9.

[51] Sabater-Vilar M, Maas RFM, Fink-Gremmels J. Mutagenicity of commercial *Monascus* fermentation products and the role of citrinin contamination. Mutat Res -Genetic Toxicol Environ Mutagenesis 1999; 444(1): 7-16.

[52] Ciegler A, Vesonder RF, Jackson LK. Production and biological-activity of patulin and citrinin from *Penicillium expansum*. Appl Environ Microbiol 1977; 33(4): 1004-6.

[53] Wang YZ, Ju XL, Zhou YG. The variability of citrinin production in *Monascus* type cultures. Food Microbiol 2005; 22(1): 145-8.

[54] Pisareva E, Savov V, Kujumdzieva A. Pigments and citrinin biosynthesis by fungi belonging to genus *Monascus*. Zeitschrift Fur Naturforschung C- J Biosci 2005; 60(1-2): 116-20.

[55] The Ministry of Health and Welfare of Japan. *Monascus* Color. Japan's Specifications and Standards for Food Additives (7th ed.), 2000. pp. Sect. D257.

[56] Kitabatake N, Trivedi AB, Doi E. Thermal-decomposition and detoxification of citrinin under various moisture conditions. J Agric Food Chem 1991; 39(12): 2240-4.

[57] Hirota M, Menta AB, Yoneyama K, Kitabatake N. A major decomposition product, citrinin H2, from citrinin on heating with moisture. Bioscol Biotechnol Biochem 2002; 66(1): 206-10.

[58] Bentrivedi A, Hirota M, Doi E, Kitabatake N. Formation of a new toxic compound, citrinin H1, from citrinin on mild heating in water. J Chem Soc Perkin 1 1993; (18): 2167-71.

[59] Su YC. Anka (Red-Koji) Products and It's Research Development in Taiwan (in Chinese). Symposium on Functional Fermentation Products. Taipei, Taiwan, 2001; pp. 67-112.

[60] Juzlova P, Martinkova L, Kren V. Secondary metabolites of the fungus *Monascus*: A review. J Ind Microbiol 1996; 16(3): 163-70.

[61] Pattangul P, Pinthong R, Phianmongkhol A, Tharatha S. Mevinolin, citrinin and pigments of adlay angkak fermented by *Monascus* sp. Int J Food Microbiol 2008; 126(1-2): 20-3.

[62] Su YC, Huang JH. Studies on the production of anka-pigment. J Chin Agric Chem Soc 1976; 14(1-2): 45-58.

[63]   Lotong N, Suwanarit P. Fermentation of Ang-Kak in plastic bags and regulation of pigmentation by initial moisture-content. J Appl Bacteriol 1990; 68(6): 565-70.

[64]   Teng SS, Feldheim W. The fermentation of rice for anka pigment production. J Ind Microbiol Biotechnol 2000; 25(3): 141-6.

[65]   Chen FS, Hu XQ. Study on red fermented rice with high concentration of monacolin K and low concentration of citrinin. Int J Food Microbiol 2005; 103(3): 331-7.

[66]   Chiu CH, Ni KH, Guu YK, Pan TM. Production of red mold rice using a modified Nagata type koji maker. Appl Microbiol Biotechnol 2006; 73(2): 297-304.

[67]   Wang JJ, Lee CL, Pan TM. Modified mutation method for screening low citrinin-producing strains of *Monascus purpureus* on rice culture. J Agric Food Chem 2004; 52(23): 6977-82.

[68]   Wang JJ, Lee CL, Pan TM. Improvement of monacolin K, gamma-aminobutyric acid and citrinin production ratio as a function of environmental conditions of *Monascus purpureus* NTU 601. J Ind Microbiol Biotechnol 2003; 30(11): 669-76.

[69]   Lee CL, Wang JJ, Kuo SL, Pan TM. *Monascus* fermentation of dioscorea for increasing the production of cholesterol-lowering agent - monacolin K and antiinflammation agent - monascin. Appl Microbiol Biotechnol 2006; 72(6): 1254-62.

[70]   Sato K, Naito I. Acids and Alcohols as nutrients for *Monascus*. J Agric Chem Soc Jpn 1935; 11: 473-9.

[71]   McHan F, Johnson GT. Zinc and Amino Acids: Important components of a medium promoting growth of *Monascus purpureus*. Mycologia 1970; 62(5): 1018-31.

[72]   Lin CF. Isolation and cultural conditions of *Monascus* sp for production of pigment in a submerged culture. J Ferment Technol 1973; 51(6): 407-14.

[73]   Lin CF, Suen SJT. Isolation of hyperpigment-productive mutants of *Monascus* sp-F-2. J Ferment Technol 1973; 51(10): 757-9.

[74]   Carels M, Shepherd D. Effect of different nitrogen-sources on pigment production and sporulation of *Monascus* species in submerged, shaken culture. Can J Microbiol 1977; 23(10): 1360-72.

[75]   Chen MH, Johns MR. Effect of pH and nitrogen-source on pigment production by *Monascus purpureus*. Appl Microbiol Biotechnol 1993; 40(1): 132-8.

[76]   Hajjaj H, Blanc P, Groussac E, Uribelarrea JL, Goma G, Loubiere P. Kinetic analysis of red pigment and citrinin production by *Monascus ruber* as a function of organic acid accumulation. Enzyme Microb Technol 2000; 27(8): 619-25.

[77]   Pereira DG, Tonso A, Kilikian BV. Effect of dissolved oxygen concentration on red pigment and citrinin production by *Monascus purpureus* ATCC 36928. Braz J Chem Eng 2008; 25(2): 247-53.

[78]   Lin TF, Demain AL. Leucine interference in the production of water-soluble red *Monascus* pigments. Arch Microbiol 1994; 162(1-2): 114-9.

[79]   Lin WY, Ting YC, Pan TM. Proteomic response to intracellular proteins of *Monascus pilosus* grown under phosphate-limited complex medium with different growth rates and pigment production. J Agric Food Chem 2007; 55(2): 467-74.

[80]   Miyake T, Mori A, Kii T, *et al.* Light effects on cell development and secondary metabolism in *Monascus*. J Ind Microbiol Biotechnol 2005; 32(3): 103-8.

[81]   Chen YP, Tseng CP, Liaw LL, *et al.* Cloning and characterization of monacolin K biosynthetic gene cluster from *Monascus pilosus*. J Agric Food Chem 2008; 56(14): 5639-46.

[82]   Mapari SAS, Hansen ME, Meyer AS, Thrane U. Computerized screening for novel producers of *Monascus*-like food pigments in *Penicillium* species. J Agric Food Chem 2008; 56(21): 9981-9.

[83]   Yang YJ, Lee I. *Agrobactrium tumefaciens*-mediated transformation of *Monascus ruber*. J Microbiol Biotechnol 2008; 18(4): 754-8.

CHAPTER 7

Fungal Biofiltration for the Elimination of Gaseous Pollutants from Air

Sergio Revah[1*], Alberto Vergara-Fernández [2] and Sergio Hernández [3]

[1]Departamento de Procesos y Tecnología, Universidad Autónoma Metropolitana-Cuajimalpa, c/o IPH, UAM-Iztapalapa, Av. San Rafael Atlixco No. 186, 09340 México D. F., México; [2]Escuela de Ingeniería Ambiental, Facultad de Ingeniería, Universidad Católica de Temuco, Rudecindo Ortega 02950, Casilla 15-D, Temuco, Chile and [3]Departamento de Ingeniería de Procesos e Hidráulica, Universidad Autónoma Metropolitana-Iztapalapa, Apdo. Postal 55-534, CP 09340, México DF, México.

Abstract: Biological technologies for air pollution control are environmentally sound and economic alternatives and are being increasingly used in industry. Biofiltration involves passing waste air through a packed reactor containing active microorganisms capable of degrading pollutants. In general, biofilters achieve the highest rates of removal for compounds that are water soluble and biodegradable.

The low solubility of hydrophobic molecules in the aqueous biofilm is one of the major problems for their treatment in biofilters. However, this obstacle may be reduced using fungi as the biological agent. Fungi have several advantages for the abatement of hydrophobic volatile organic compounds (VOCs) in gas-phase biofilters, including their ability to degrade a large number different molecules, their resistance to low humidity and pH, their capacity to colonize unoccupied space with the aerial hyphae and to penetrate the solid support increasing the availability of nutrients. On the other hand, fungi grow slower than bacteria, their filamentous growth promotes increased pressure drop and under some conditions may produce spores that could represent some health hazard if not contained. This review will present the state of the art on the research being done with fungal biofiltration, covering the diversity of molecules and fungal species studied, the characteristics of fungal growth in biofilters, the type of reactors and supports used, the operational problems and mathematical modeling of these processes.

INTRODUCTION

Biological air treatment systems provide a suitable low-cost and environmentally friendly alternative to traditional and energy-intensive methods such as incineration, absorption, condensation and adsorption. In biofilters, the polluted air is forced through a packed bed on which microorganisms are attached. Biodegradable volatile compounds and oxygen diffuse into the biofilm where they are subsequently transformed by a diverse microbial population into CO_2, H_2O and biomass. This is now a popular technology for the treatment of odors from diverse sources and some industrial emissions [1, 2].

Biofiltration processes usually contain mixed microbial populations adapted for the particular pollutant, type of equipment and environmental conditions (pH, temperature, humidity, etc) and contain mainly bacteria. As these systems do not operate aseptically, they are open to incoming microorganisms which can establish, under certain conditions within the microbial consortium.

It is common to find filamentous fungi colonizing the surface of the supports and developing aerial growth. Recently, some of these fungi have been isolated and inoculated to biofilters to preferentially foster their growth and activity.

The use of fungi in biofilters may present, at least theoretically, several advantages when compared with bacterial ones such as: enzymatic diversity and grow under environmentally stressed conditions such as low nutrient availability, low water activity and low pH values, where bacterial growth might be limited [3]. Also, some authors have suggested the occupation of free space by the formation of aerial hyphae and its surface hydrophobicity may present some advantages, especially with sparsely soluble gases [4, 5]. Despite the various benefits of the use of fungi in biofilters, these systems have several shortcomings that will be discussed later and been poorly explored at the pilot scale and most studies have been performed at laboratory scale.

*Address correspondence to Sergio Revah:** Departamento de Procesos y Tecnología, Universidad Autónoma Metropolitana-Cuajimalpa, c/o IPH, UAM-Iztapalapa, Av. San Rafael Atlixco No. 186, 09340 México D. F., México; E-mail: srevah@xanum.uam.mx

Ana Lúcia Leitão (Ed)

The present chapter presents the state of the art on the research being done with fungal biofiltration. The aspects covered include the diversity of molecules and fungal species studied, the characteristics of fungal growth in biofilters, the type of reactors and supports used and the operational conditions. Also, a discussion of the mathematical models used in biofilters to describe the growth of the fungus is presented.

TECHNOLOGY FOR VOCS AND ODOR CONTROL

Today there are a variety of technologies available for treatment of VOCs emissions and odors. To select an appropriate control method it is important to consider their physical, thermodynamic and reaction properties. The existent technologies can be classified in physical, chemical or biological methods. Generally, the physical processes are applied for waste gas streams with high pollutant concentration or with extreme pressure or temperature conditions. Important parameters for biological treatment are the concentration, the solubility and the biodegradability of the compounds [1, 2].

The interest in the use of biofiltration is based on its low energy requirements; it does not require extreme conditions of operation, is conducted at normal temperatures (10–40 ºC) and atmospheric pressure. The pollutant is destroyed and not just transferred to another phase. Biological methods are effective, robust, simple to operate and ecologically friendly as compared with most physicochemical treatments. Generally, investment and operating costs are also lower which is important when a technology is chosen [2, 6].

Biofiltration is based on the capability of microorganisms to transform certain organic and inorganic pollutants into less toxic and odorless compounds. As the pollutants are in air, the processes of microbial degradation are generally oxidative in nature and the end products are carbon dioxide, water, sulfate, and nitrate. This technology is not a new idea; the first reports are related to the treatments of odors in soil beds from sewage [7]. During the years 1960–1970 it was shown that the biological reaction was the main mechanism for the pollutants elimination, and the first commercial applications were established in The Netherlands, West Germany and the United States [8]. The initial applications were generally open biofilters, and they were used to treat odors from a variety of sources such as sewage treatment plants, rendering facilities, compost, food processing, and farms [7].

Table 1: Description of the Three Reactor Types: Biofilter, Biotrickling Filter and Bioscrubber.

Reactor	Definition	Biomass	Liquid phase
Biofilter	A packed single reactor that is either open or closed. The packing is a natural material such as compost, peat, or soil. Additives may be added to increase the porosity. This support provides the microorganisms with nutrients and buffer capacity. The biofilter is supplied with a humid gas phase that is forced through the packing. Water and nutrients are added intermittently to keep humidity and favor microbial growth.	Fixed on a support	Fixed on a support
Biotrickling filter	Consists of vessels packed with non degradable material, usually structured or random plastics, which allow the development of a microbial film increasing the volumetric cell density. Water or medium are continually recirculated through the packing.	Fixed on a support	Flowing
Bioscrubber	It consists of two vessels: an absorption tower and a biological reactor. In the absorber the pollutants are transferred to an aqueous medium to near saturation conditions. The liquid is recirculated either concurrent or countercurrent with the gas.	Suspended	Flowing

Performance is generally reported as elimination capacity, (Eq. 1) defined as:

$$EC = \frac{Q(C_{gi} - C_{go})}{V_r}$$

(eq. 1)

where EC is the elimination capacity (g pollutant $m^{-3}_{reactor}$ min^{-1}), Q is the inlet airflow (m^3 min^{-1}), C_{gi} is the inlet concentration of pollutant (g pollutant m^{-3}), C_{go} is the outlet pollutant concentration (g pollutant m^{-3}). Further developments included the use of better supports, allowing improved performance and stability. Developments in biofilter configurations included closed systems and improved control. Since the '80s, there has been extensive research in the fundamentals of biological air treatment systems, and new have been successfully implemented, with the aim to increase biodegradation rates and efficiency [9].

Although the basic mechanisms are the same for all biological air treatment systems, there are three different configurations to achieve transfer and biodegradation. These systems are: biofilters; biotrickling filters and bioscrubbers and are briefly described in Table **1** [1, 2, 10].

TECHNOLOGICAL ASPECTS TO BE CONSIDER IN BIOFILTRATION

The main features that a gaseous pollutant must have for its successful treatment in a biological system are their high biodegradability and low toxicity. Pollutants with low molecular weight, high solubility and a simple chemical structure are more efficiently eliminated. The compounds with more complex structure require more energy to be degraded [1, 2]. Table **2** shows some of the compounds that have been removed by biological treatment systems.

Table 2: Examples of Compounds Treated by Biofiltration Systems.

Ketone	Formaldehyde
Ammonium	Hexane
Benzene	Methanol
Butanol	Methyl ethyl ketone
Dimethyl disulfide	Methane
Diethyl ether	Methyl mercaptan
Dimethyl sulfide	Methyl tert-butyl ether
Nitrogen dioxide	Nitrogen oxide
Ethanol	Pentane
Ethylbenzene	Hydrogen sulfide
Styrene	Toluene
Phenol	Xylene

Adapted from references [11, 12].

Microorganisms

The microorganisms involved in the degradation of pollutants in biofilters include bacteria, actinomycetes and fungi [13]. Most of the microbial consortia in biofilters consist of heterotrophic microorganisms although autotrophic species are dominant for elimination of inorganic pollutants such as hydrogen sulfide or ammonia [14]. Bacteria are the dominant organisms for VOCs elimination and conditions favoring their activity, such as high humidity and neutral pH are promoted [1, 15]. The selection of the microbial population is determined by the nature of the pollutant [16]. In general, naturally produced substances, such as those found in foul odors from water treatment plants, composting facilities, *etc.,* are easily biodegradable at low concentrations. They can be treated with the natural population found in biofilter media (such as peat, compost, bark, wood shavings, *etc.),* or can be inoculated with a diverse mixed population, such as that found in activated sludge. Appropriate elimination capacity and efficiency require active and sufficient biomass, so inoculation from an adapted population is desirable. The same startup approach can be used for industrial applications when the pollutant has been shown to be sparsely biodegradable, where a less diverse population is established. There are extreme cases of substances that are difficult to degrade so the reactor has to be inoculated with specialized strains or consortia to initiate degradation [1, 2]. In the last decade, several studies have reported biofiltration using pure cultures and microbial consortia [10, 17-21].

Moisture Content

The humidity in the biofilters is a critical step to maintain a proper performance, as biological activity is highly dependent on water activity (Aw). The heat generated by the biological reaction and the humidity of the incoming air determines the rate of water loss and requirements for water restoration [22].

Biotrickling filters and bioscrubbers have a mobile liquid phase, therefore, they do not have moisture problems during operation. Biofilters (fixed bed) on the other hand, are generally packed with natural supports such as compost, wood shavings, bark, peat moss [1,2], and consequently have a static liquid phase within the support and require intermittent water addition and air pre-humidification systems to maintain the biological activity [6]. Drying of the bed has reported to be the single most important problem in biofiltration malfunctioning and strong water losses have been observed in equipments with no proper humidification of gas stream and high rates of pollutant elimination. On the other hand, excessive water content can reduce the contact area between the air and the biofilm, creating anaerobic zones within the reactor as a result of the difficulty related to oxygen transfer.

pH

The pH conditions required in the operation of biological treatment system gases are generally set in the range of 6 to 8 to favor the biological activity of heterotrophic bacteria [6]. However, there are many biological processes that generate acidic, basic or inhibitory compounds, as reported in the treatment of chlorinated compounds, hydrogen sulfide, methyl sulfide, ammonia, among others [23-26]. However, under acidic conditions in a range of 3.0 and 4.5, for some VOCs high removal efficiencies have been obtained as a result of fungal growth [27, 28].

Solubility of Gaseous Pollutants

The amount of gaseous pollutant that can be solubilized in the biofilm at equilibrium for gas-biomass systems at low concentrations can be represented by the partition coefficient, which corresponds to an apparent Henry's constant, represented by Henry's law (Eq. **2**).

$$C_{gi} = H_i Cl_i$$

(eq. 2)

Where, C_{gi} is the concentration of pollutant "i" in the gas phase, H_i is the Henry's constant, and Cl_i is the concentration of "i" in the liquid/biotic phase. Henry's constant can be found in the literature in different units. Using a non-dimensional Henry's constant (mg gas L^{-1} over mg liq L^{-1}), substances with values over 0.01 are considered volatile, and the higher the value, the less soluble the substance is in water. Henry's constant depends on temperature and the chemical potential of the liquid phase. Some values for water are shown in Table **3**.

In general, the presence of biomass, extracellular polymers and products in the liquid phase improves the solubility of hydrophobic compounds [29]. This feature is accentuated when using filamentous fungi, due to their hydrophobic characteristics [5].

Table 3: Henry's Constant for some Common Compounds at 25 °C.

Compound	Henry's constant (non-dimensional)
Hexane	30.9
Oxygen	29.1
Hydrogen sulfide	0.92
Toluene	0.25
Benzene	0.22
MIBK (methyl iso-butyl ketone)	0.016
Ethanol	0.0012
Ammonia	0.0005

From reference [2].

The transport of a gaseous compound or a mixture from air to an aqueous phase is often the limiting process in a biofiltration system, especially if they are in low concentrations, as generally found in odorous pollutants or when they have low solubility (high Henry's constant). Biotrickling filters have been traditionally used to treat pollutants with high solubility due to presence of a recirculating liquid. On the other hand, traditional biofilters with natural packing have been preferred for less soluble pollutants [16, 30]. Other studies have also suggested that the transport of gaseous substrates to a biofilm may not be limited by the liquid phase, and direct mass transfer to the unsaturated biofilm may occur [31, 32].

Deshusses [16] investigated the elimination capacity of 18 VOCs, with a broad range of Henry's constant using biofilters for 48 hours for each compounds. They concluded that biodegradation of these VOCs in biofilters was strongly influenced by its availability or the Henry's constants. Another important aspect related to the effect of the Henry's constant is the oxygen limitation (O_2), which can occur in biofilters used to remove VOCs with low values of Henry's constant (hydrophilic compounds), given their greater presence in the aqueous phase and biofilm. Kirchner [33] found the range of oxygen diffusion rate in the biofilm where there was no limitation by varying the oxygen content of the inlet gas in the treatment of two hydrophilic VOCs (acetone and propionaldehyde).

FUNGAL BIOFILTRATION

Although bacteria are the predominant organism in most of the reported applications for VOCs elimination, the presence of fungi, including yeasts, in biofilters has been reported frequently even in cases where the conditions were manipulated to avoid their growth [19, 34]. Fungi are widely distributed in nature and have a large diversity of morphology, physiology and biological activity. In biofilters, Fig. **1**, their mycelial growth integrates in complex networks occupying the free aerial space and colonizing the supports where they uptake required water soluble nutrients such as N, S, Mg, etc.

Reports of fungi isolated from biofilters have shown that they can metabolize hydrocarbons (aliphatic and aromatic) as their sole source of carbon and energy. The hyphae produced by some species have a very large surface area compared to its volume, which facilitates the diffusion of nutrients. Wösten [35] suggests that the aerial mycelium, which is in direct contact with the gas, could facilitate the degradation of hydrophobic compounds as compared to the bacterial biofilms.

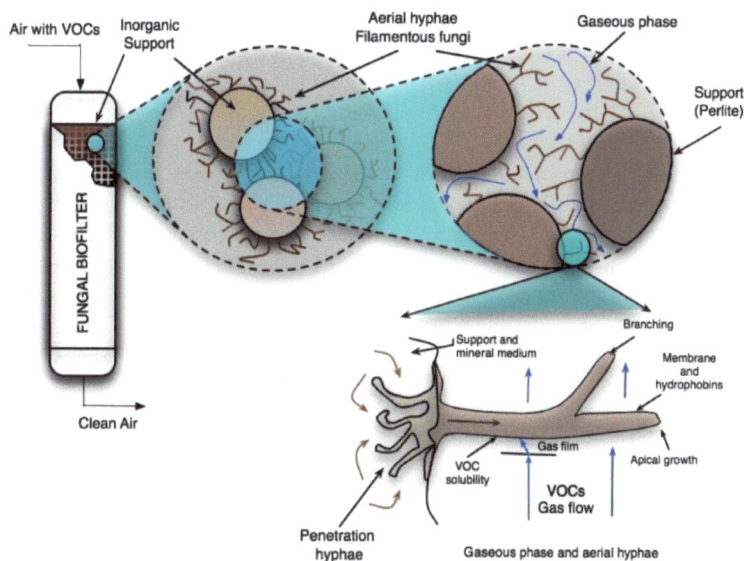

Figure 1: Representation of fungal growth in biofilters.

The optimal temperature range for microbial growth is between 20 and 45 °C and is different for each species. Most fungi are mesophilic, growing in temperature ranges between 10 and 35 °C [36]. In the case of bacteria, the range of water activity for growth is 1.00 to 0.90 [37]. In contrast, fungi can tolerate a smaller range between 0.75-0.65 [38]. In a bacterial biofilter, the relative moisture of inlet gas must be near 100% to prevent bed drying and maintain microbial activity [22]. For fungi, it has been reported that relative humidity near 70% can support growth [36]. Several investigations suggest that fungi can degrade a wide variety of VOCs at elimination capacities (EC) equal or greater than those observed in bacterial systems [27, 39, 40].

In one of the first reports on fungal biofiltration, van Groenestijn [41] reported a biofilter containing fungi capable of degrading styrene, which achieved an EC between 50 and 110 g m^{-3} h^{-1}. These values were higher than those obtained in bacterial biofilters. For toluene García-Peña [27], Woertz [40] and Aizpuru [51] have reported EC of 260, 270 and 290 g m^{-3} h^{-1} using *Paecilomyces variotii (previously reported as Scedosporium apiospermum)* and *Exophiala lecanii-corni*. These EC values were significantly higher than previously reported with bacteria where values up to 50 g m^{-3} h^{-1} were generally found. While, van Groenestijn [4] found that for α-pinene biofiltration based on the action of fungi exhibited higher EC than those reported with other microbial populations. Moe [53] studied the elimination of a VOC mixture composed by *n*-butil acetate, methyl ethyl ketone, methyl ethyl propyl ketone and toluene, using a culture of several species of fungi, including *Cladosporium sphaerospermun, Exophilia jenselmei, Fusarium oxysporum* and *Fusarium nygamai,* in this case the average EC$_{total VOC}$ was of 92.1 g m^{-3} h^{-1}. On the other hand, Spigno [18, 54] obtained an average EC of 150 g m^{-3} h^{-1} working on a biofilter, inoculated with the fungus *Aspergillus niger* to eliminate *n*-hexane. Arriaga [3, 56, 74] also studied the biodegradation of *n*-hexane in biofilters inoculated with *Fusarium solani*, obtaining maximum EC between 150 and 180 g m^{-3} h^{-1}.

Contrary to bacterial degradation, there is much less information on the fungal mechanisms. Biodegradation pathways for benzene, toluene, ethylbenzene and xylene (BTEX) in pure cultures have been reported by several authors [42-46]. These studies indicated that the alkyl-benzenes (toluene, ethylbenzene and xylenes) were degraded by a combination of assimilation and cometabolism by the same monooxygenase, similar to the bacteria pathway [45]. A more detailed study with *P. variotti* performed by García-Peña [47] concluded that toluene biodegradation occurs both through hydroxylation of a methyl group and by the *p*-cresol pathway.

Several studies have reported the successful degradation of VOCs in microcosms and biofilters with fungi at laboratory scale; they are summarized in Table **4**. In the same Table, there are several species of fungi that have been identified as able to degrade VOCs in liquid medium. Some of the compounds that have been removed individually in biofilters include: styrene, toluene, methyl ethyl ketone, α-pinene, *n*-hexane, mixtures of BTEX and ethanol.

As shown in Table **4**, most papers report biofiltration of gases with a single compound, and only a few deal with the treatment of mixtures of VOCs. Recent reports show that pure and mixed fungal cultures may be appropriate for the treatment of mixtures of compounds such as BTEX and hydrophobic compounds (pentane and hexane) [52, 62]. Also, this Table includes some liquid media cultivation with hydrophilic compounds related to biofiltration studies (i. e. ethanol, phenol and MTBE*).*In those studies, the mixtures of compounds are of the same chemical family, and the objective was to increase biofilter degradation efficiency through the optimization of operating conditions. The treatment of complex mixtures can be problematic when using biological methods. Slow degradation rates in some cases suggest inhibition phenomena and catabolic repression [1, 63-65]. Most of the reports with fungal biofilters use axenic cultures, but the possibility of enriching local fungi by maintaining proper conditions such as low pH and moisture content and also by intermittently adding bacterial antibiotics has been successfully achieved [3].

Table 4: Species of Fungi Reported that Degrade VOCs in Biofilter and Liquid Medium.

Fungi	Compounds	System	Reference
Phanerochaete Chrysosporium	Mixture of BTEX	Bottles with pure cultures	[39, 42]
Bjerkandera adusta	Styrene, TEX	Bottles with pure cultures	[39]
Trametes versicolor	Styrene, TEX	Bottles with pure cultures	[39]
Pleurotus ostreatus	Styrene, TEX	Bottles with pure cultures	[39]
Cladosporium Sphaerosporium	T	In bottles, culture isolated from compost biofilter	[43]
Exophiala jeanselmei	Styrene T	Biofilter inoculated with pure culture isolated from a biofilter	[48, 49] [40]
Fusarium oxysporum	MEK	Isolated from a biofilm reactor and tested in liquid medium	[34]
Geotrichum	MEK	Isolated from a biofilm reactor and tested in	[34]

candidum		liquid medium	
Exophiala sp.	TE	Test in bottles cultures obtained from collections	[45]
Leptodontium sp.	T	Test in bottles cultures obtained from collections	[45]
Pseudeurotium zonatum	T	Test in bottles cultures obtained from collections	[45]
Paecilomyvces variotti (formerly Scedosporium apiospermum)	T	Biofilter inoculated with pure culture isolated from a biofilter	[27]
Paecilomyces variotti	T BTEX	Biofilter Biofilter	[50, 51, 52]
Cladophialophora sp.	Mixture of styrene, TEX	In bottles with pure cultures, isolated from soil contaminated with BTEX	[46]
Exophiala oligosperma	T	Biofilter	[50]
Cladosporium sphaerospermun	T, *n*-butyl acetate, MEK	Biofilter	[53]
Fusarium oxysporum	T, *n*-butyl acetate MEK	Biofilter	[53]
Fusarium nygamai	T, *n*-butyl acetate MEK	Biofilter	[53]
Exophilia jenselmei	*n*-butyl acetate MEK	Biofilter	[53]
Aspergillus niger	*n*-hexane	Biofilter	[18] [54]
Ophiostoma species	α-pinene	Biofilter	[55]
Fusarium solani	*n*-hexane Methyl *tert*-butyl ether, ter-butyl alcohol	Biofilter	[56, 57] [58]
Rhinocladiella similis	*n*-hexane, T, ethanol, phenol	Biofilter	[59] [61]
Paecilomyces lilacinus	T	Biofilter	[60]
Candida utilis	Ethanol	Biofilter	[60]
Non axenic mixed fungi	*n*-hexane	Biofilter	[3,62]

Abbreviations: Benzene (B), Toluene (T), Ethylbenzene (E), Xylene (X), Methyl Ethyl Ketone MEK

Low VOCs solubility is a major constraint to their treatment in biofilters because of the presence of liquid film covering the biofilm. However, this can be facilitated with the use of fungi as the biological agent as they are more resistant, than bacteria, to acid and low moisture conditions. Furthermore, fungi facilitate the transport of hydrophobic compounds by both the formation of aerial mycelia, which has an extensive surface, and by the hydrophobic characteristics of the hyphae [12, 35]. Table **5** shows some of the advantages and disadvantages of the use of fungi in biofilters.

Low VOCs solubility is a major constraint to their treatment in biofilters because of the presence of liquid film covering the biofilm. However, this can be facilitated with the use of fungi as the biological agent as they are more resistant, than bacteria, to acid and low moisture conditions. Furthermore, fungi facilitate the transport of hydrophobic compounds by both the formation of aerial mycelia, which has an extensive surface, and by the hydrophobic characteristics of the hyphae [12, 35]. Table **5** shows some of the advantages and disadvantages of the use of fungi in biofilters.

Table 5: Advantages and Disadvantages of the Use of Fungi in Biofiltration.

Advantages	Disadvantages
Wide variety of enzymes (hydrolytic, oxidation)	Slower growth than bacteria
Degradation of an ample variety of organic compounds	Increased pressure drop
Resistance to low water activity and pH	Possible production and emission of spores
Ability to colonize the empty void (aerial hyphae)	Possible obstruction of air passage and channeling
Ability to penetrate the solid support increasing nutrient availability	Possible pathogenic

Recently, Vergara-Fernández [5] found that for the filamentous fungus *Fusarium solani* the surface hydrophobicity, fat content and partition coefficient were affected by the culture conditions (solid or liquid) and the type of carbon source. The growth of *Fusarium solani* in a more hydrophobic carbon source increases the superficial hydrophobicity and the solubility of VOCs in biomass. These changes in the fungus were not only related to its fat content, but possibly also to the presence of various types of molecules in the fungal wall including, among others hydrophobins and lipoproteins. This study also established the improvement in the partition coefficient (decrease of partition coefficient of 200 times in the fungus in comparison with water) and the largest area of gas exchange provided by the fungal growth in a biofiltration system, which increased the bioavailability of hydrophobic VOCs and thus increasing the elimination rate. The presence of hydrophobins and their effects on removal capacity and the solubility of hydrophobic VOCs biofiltration conditions was determined by Vigueras [59, 60] using the fungi *Paecilomyces lilacinus* and *Rhinocladiella similis,* respectively.

The following photo (Fig. **2**) shows laboratory-scale fungal biofilters used for the removal of hydrophobic VOCs with the fungus *Fusarium solani*, and the colonization of the support.

Figure 2: Laboratory-scale fungal biofilters inoculate with *Fusarium solani* for the *n*-hexane elimination.

Although interest in the study of fungal biofiltration has focused on the elimination of hydrophobic compounds with filamentous organisms, another area of interest that has been scarcely explored is the degradation of hydrophilic compounds using non-filamentous fungi. An example of this biofiltration system is the degradation of ethanol, and subproducts, such as, acetaldehyde, acetic acid and ethyl acetate by *Candida utilis* [61]. Where, the possibility of using the yeast-enriched support as a single cell protein source was also explored.

The limitations for the more extended use of fungal biofilters include the fact that they generally grow slower and have lower metabolic activity than bacteria [47] which may have some inconvenient on start-up where a sufficient biomass is required to obtain high removal rates. Also, as a result of the mycelial growth, obstruction of air passage may lead to increased pressure drop and possibly flow obstruction, bed compaction and reduction in the permeability has been also observed [66]. Increment in pressure drop has been reported in the operation of biofilters, in ranges between 5 to 40 and 12 to 50 mmH_2O/m bed, for 8 and 50 operation days, respectively. This behavior was related to the growth of the fungus in the biofilter [27, 57, 67].

There is extended concern on the spore emission and possible pathogenicity of the strains used but there are some authors that sustain that many fungi are not pathogenic at all [12, 67]. Furthermore, there have been several reports with industrial biofilters that show that there is a very limited spore emission and in some cases there may be even a reduction due to the filtering capacity of the biofilters [68]. However, more research is required to assure the safety for predominantly fungal system.

MODELING FUNGAL BIOFILTERS

Mathematical modeling is necessary to understand and predict the performance and scaling up of bioreactors. In 1983, Ottengraff [9] presented the first VOCs biotreatment in biofilters with a diffusion and biodegradation model of pollutants in biofilms. While several models, with increased degree of complexity were reported thereafter [1], the original description often has been used to represent biofiltration data [69]. Due to the complexity of the system, given the diversity of microbial populations and the inherent heterogeneity of the system, fixed bed biofilters are difficult to model. The modeling of these systems involve physical and biochemical phenomena, including fluid flow and diffusion properties of microbial communities and material (solid support), prediction of active area and biofilm thickness [70, 71].

The models previously mentioned represent generally the biofilm as a flat surface, typical of bacteria, and do not describe the characteristic fungal growth as seen in Fig. 1. On the other hand, there has been a lot of work describing batch fungal growth [72, 73].

For the case of modeling fungal biofilters, Spigno [18, 54] performed a simple model of axial dispersion in steady state to evaluate the *n*-hexane elimination in a biofilter using the fungus *Aspergillus niger*. This model performs the same considerations that the models used for microbial consortia and bacterial biofilter, working with a homogeneous biofilm and constant biomass. An improvement including actual partition coefficient in the fungi, biofilm thickness, superficial area and effective diffusivity has been recently presented by Arriaga [74].

In general, the latter models are based on assumption of a biofilm as a pseudo-homogeneous phase. However, in the case of aerial biofilms, such as those generated by filamentous fungi, these do not provide information of inter-particle phenomena occurring in the biofilter. The aerial growth in filamentous microorganisms, allows covering the empty void between the solid supports in the biofilter, which is completely different situation to that occurring in bacterial biofilms where growth induces the increase in the depth of the biofilm, which brings about reduced transport and even anaerobic layers. Furthermore, the hydrophobic surface of the fungus and its resistance to drought allow direct absorption from the air pollutant. Furthermore, the increase in area produced by filamentous fungi may contribute to increased mass transfer to be used in biofilters, according to Braun-Lullemann [39].

Vergara-Fernández [57] developed a mathematical model considering the physical and biological phenomena in a fungal biofilter. The biofilter is mathematically described and the main physical (mass transfer, partition and transport area), kinetic data (substrate inhibition and affinity, growth and degradation rates and maintenance coefficient), and morphological parameters of aerial hyphae (hyphal diameter, average length of segments, critical length, maximum length of distal individual hyphae, branching number in individual hyphae, colony rate extension) were obtained by independent experiments for model verification. The model proposed in this study [57] describes the increase in the transport area by the growth of the filamentous cylindrical mycelia and its relation with *n*-hexane elimination in quasi-stationary state in a biofilter. In order to mathematically describe the system, four processes were considered: (1) mass transfer of VOCs in the bulk gas, (2) mass transfer of VOCs into the gas layer around the mycelium and considering the experimentally obtained gas-biomass partition coefficient, (3) mass transfer and uptake of the nitrogen source through the elongating mycelia, (4) and the kinetic of mycelial growth. Processes (2) and (3) include movable boundary conditions to account for the mycelial growth. The model describing fungal growth includes Monod-Haldane kinetic and hyphal elongation and ramification.

CONCLUSIONS

Fungal biofilters have shown to be very efficient for the treatment of hydrophobic compounds. Although there are an increasing number of publications covering a wider range of pollutants and conditions, the technology still needs to be evaluated at bigger scale and for longer operation times. The safety of fungal biofilters needs also to be addressed before industrial applications are developed.

ACKNOWLEDGEMENTS

The authors acknowledge the support of the International Scientific Cooperation Program CONICYT-Chile /CONACYT – México

REFERENCES

[1]   Devinny JS, Deshusses MA, Webster TS. Biofiltration for air pollution control. Lewis Publishers: Boca Raton, FL 1999.

[2]   Revah S, Morgan-Sagastume JM. In: Shareefdeen Z, Singh A, Eds. Methods for odor and VOC control in Biotechnology for odour and air pollution. Heidelberg: Springer-Verlag 2005; pp. 29-64.

[3]   Arriaga S, Revah S. Improving hexane removal by enhancing fungal development in a microbial consortium biofilter. Biotechnol Bioeng 2005; 90: 107-15.

[4]   van Groenestijn JW, Liu JX. Removal of alpha-pinene from gases using biofilters containing fungi. Atmos Environ 2002; 36(35): 5501-8.

[5]   Vergara-Fernandez A, Van Haaren B, Revah S. Phase partitioning of gaseous surface hydrophobicity of *Fusarium solani* when grown in liquid and solid media with hexanol and hexane. Biotechnol Lett 2006; 28: 2011-7.

[6]   Baltzis BC, Biofiltration of VOC Vapors. In: Lewandowski GA, DeFilippi LJ, Eds. Biological treatment of hazardous waste. New York: Wiley Interscience, 1998; pp. 119-50.

[7]   Leson G, Winer AM. Biofiltration: an innovative air pollution control technology for VOC emissions. Air Waste 1991; 41:1045-54.

[8]   Ottengraf SPP. Exhaust gas purification. In: Rehm H-J, Reed G, Eds. Biotechnology. Weinheim: VCH Verlagsgesellschaft 1986; pp. 426-52.

[9]   Ottengraf SPP, van den Oever AHC. Kinetics of organic compound removal from waste gases with a biological filter. Biotechnol Bioeng 1983; 25: 3089-102.

[10]   Schroeder ED. Trends in application of gas-phase bioreactors. Rev Environ Sci Biotechnol 2002; 1:65-74.

[11]   Swanson WJ, Loehr RC. Biofiltration: fundamentals, design and operations principles, and applications. J Environ Eng 1997; 538-46.

[12]   Kennes C, Veiga MC. Fungal biocatalysts in the biofiltration of VOC-polluted air. J Biotechnol 2004; 113: 305-19.

[13]   Deshusses MA. Biological waste air treatment in biofilters. Curr Opin Biotechnol 1997; 8: 335-9.

[14]   Kircher K, Schlachter U, Rehm HJ. Biological purification of exhaust air using fixed bacterial monocultures. Appl Microbiol Biotechnol 1989; 31: 629-32.

[15]   van Groenestijin JW, de Bont JAM. Isolation and characterization of fungi growing on volatile aromatic hydrocarbons as their sole carbon and energy source. Mycol Res 2001; 105 (4): 477-84.

[16]   Deshusses MA, Johnson CT. Development and validation of a simple protocol to rapidly determine the performance of biofilters for VOC treatment. Environ Sci Technol 2000; 34: 461-7.

[17]   Kleinheinz GT, Bagley ST. Biofiltration for the removal and "detoxification" of a complex mixture of volatile organic compounds. J Ind Microbiol Biotechnol 1998; 20: 101-8.

[18]   Spigno G, Pagella C, Fumi MD, Molteni R, De Faveri DM. VOCs removal from waste gases: gas-phase bioreactor for the abatement of hexane by *Aspergillus niger*. Chem Eng Sci 2003; 58:739-46.

[19]   Auria R, Frere G, Morales M, Acuña ME, Revah S. Influence of mixing and water addition on the removal rate of toluene vapors in a biofilter. Biotechnol Bioeng 2000; 68: 448-55.

[20]   Vinarov AY, Robysheva ZN, Smirnov VN, Sokolov DP. Studies of the stability of microbial association use in industrial biofiltering of gaseous discharges. Appl Biochem Microbiol 2003; 38(5): 445-9.

[21]   Choi SC, Oh YS. Simultaneous removal of benzene, toluene and xylenes mixture by a constructed microbial consortium during biofiltration. Biotechnol Lett 2002; 24: 1269-75.

[22]   Morales M, Hernández S, Cornabé T, Revah S, Auria R. Effect of drying on biofilter performance: modeling and experimental approach. Environ Sci Technol 2003; 37: 985-92.

[23]	Shoda M, Methods for the biological treatment of exhaust gas. In: Marthin M, Ed. Biological degradation of wastes. New York: Elsevier Science Publishers 1991; pp. 31-46.

[24]	Ergas S, Kinney K, Fuller M, Scow K. Characterization of a compost biofiltration system degrading dichloromethane. Biotechnol Bioeng 1994; 44: 1048-54.

[25]	Kennes C, Huub H, Cox J, Doddema HJ, Harder W. Design and performance of biofilters for the removal of alkylbenzene vapors. J Chem Technol Biotechnol 1996; 66: 300-4.

[26]	Chung Y, Huang C, Tseng C. Removal of hydrogen sulphide by immobilized *Thiobacillus* sp. strain CH11 in a biofilter. J Chem Technol Biotechnol 1997; 69: 58-62.

[27]	García-Peña EI, Hernández S, Favela-Torres E, Auria R, Revah S. Toluene biofiltration by the fungus *Scedosporium apiospermun* TB1. Biotechnol Bioeng 2001; 76(1): 61-9.

[28]	Qi B, Moe WM. Performance of low pH biofilters treating a paint solvent mixture: Continuous and intermittent loading. J. Hazardous Mater 2006; B135: 303-310.

[29]	Davison BH, Barton JW, Klasson KT, Francisco AB. Influence of high biomass concentrations on alkane solubilities. Biotechnol Bioeng 2000; 68(3): 279-84.

[30]	De Header B, Overmeire A, Van Langenhove H, Verstraete W. Ethene removal from a synthetic waste gas using a dry biobed. Biotechnol Bioeng 1994; 44: 642-8.

[31]	Rihn MJ, Zhu X, Suidan MT, Kim BJ, Kim BR. The effect of nitrate on VOC removal in the trickle bed biofilters. Water Res 1997; 31(12): 2997-3008.

[32]	Zhu X, Suidan MT, Alonso C, Yu T, Kim BJ, Kim BR. Biofilm structure and mass transfer in a gas phase trickled bed biofilter. Water Sci Technol 2001; 43(1): 285-93.

[33]	Kirchner K, Wagner S, Rehm HJ. Exhaust gas purification using biocatalysts (fixed bacteria monocultures) the influence of biofilm diffusion rate (O_2) on the overall reaction rate. Appl Microbiol Biotechnol 1992; 37: 277-9.

[34]	Agathos SN, Hellin E, Ali-Khodja H, Deseveaux S, Vandermesse F, Naveau H. Gas-phase methyl ethyl ketone biodegradation in a tubular biofilm reactor: microbiological and bioprocess aspects. Biodegradation 1997; 8: 251-64.

[35]	Wösten HAB, de Vocht ML. Hydrophobins, the fungal coat unraveled. Biochim Biophys Acta 2000; 1469:79-86.

[36]	Deacon JW. Modern Mycology. Blackwell Scientific Ltd: Oxford. 1997.

[37]	Beuchat LR. Influence of water activity on growth, metabolic activities, and survival of yeast and molds. J Food Prot 1983; 46: 135-44.

[38]	Jay JM. Modern Food Microbiology, Chapman & Hall International Thomson Publishing: New York 1992.

[39]	Braun-Lüllemann A, Johannes C, Majcherczyk, Hüttermann A. The use of white-rot fungi as active biofilters. In: Heichee RE, Sayles RD, Skeen RS, Eds. Biological Unit Processes for Hazardous Waste Treatment: Battelle Press 1995; pp. 235-40.

[40]	Woertz JR, Kinney KA, Szaniszlo PJ. A fungal vapor-phase bioreactor for the removal of nitric oxide from waste gas streams. Air Waste 2001; 51: 895-902.

[41]	van Groenestijn, Harkes JW, Cox HM, Doddema H. Ceramic materials in biofiltration. *Proceedings of the Conference on Biofiltration*, Los Angeles, California, October 5-6. 1995

[42]	Yadav JS, Reddy CA. Degradation of benzene, toluene, ethylbenzene and xylenes by the lignin-degrading basidiomycete *Phanerochaete chrysoporium*. Appl Environ Microbiol 1993; 59: 756-62.

[43]	Weber FJ, Hage KC, DeBont JAM. Growth of the fungus *Clodosporium shphaerospermum* with toluene as the sole carbon and energy source. Appl Environ Microbiol 1995; 61: 3562-6.

[44]	Tsao CW, Song HG, Bartha R. Metabolism of benzene, toluene, and xylene hydrocarbons in soil. Appl Environ Microbiol 1998; 64(12): 4924-9.

[45]	Prenafeta-Boldu FX; Kuhn A, Luykx DMAM; Anke H; van Groenestijin JW, de Bont, JAM. Isolation and characterization of fungi growing on volatile aromatic hydrocarbons as their sole carbon and energy source. Mycol Res 2001; 105 (4): 477-84.

[46]	Prenafeta-Boldu FX, Vervoort J, Grotenhuis TC, van Groenestijn JW. Substrate interactions during the biodegradation of benzene, toluene, ethylbenzene, and xylene (BTEX) hydrocarbons by the fungus *Cladophialophora* sp. Strain T1. Appl Environ Microbiol 2002; 68: 2660-5.

[47]	García-Peña I, Hernández S, Auria R, Revah S. Correlation between biological activity and reactor performance in the biofiltration of toluene with the fungus *Paecilomyces variotii* CBS 115145. Appl Environ Microbiol 2005; 71: 4280-5

[48]	Cox HHJ, Magielsen FJ, Doddema HJ, Harder W. Influence of the water content and water activity on styrene degradation by *Exophiala jeanselmei* in biofilters. Appl Microbiol Biotechnol 1996; 45:851-6.

[49]	Cox HHJ, Moerman RE, van Baalen S, van Heiningen WNM, Doddema HJ, Harder W. Performance of a styrene-degrading biofilter containing the yeast *Exophiala jeanselmei*. Biotechnol Bioeng 1997; 53: 260-6.

[50]   Estévez E, Veiga MC, Kennes C. Fungal biodegradation of toluene in gas-phase biofilters. In: Verstraete, W. (Ed.), Proceedings of the 5th European Symposium on Environmental Biotechnology. Oostende: Belgium 2004; pp. 337-40.

[51]   Aizpuru A, Dunat B, Christen P, Auria R, García-Peña I, Revah S. Fungal biofiltration of toluene on ceramic rings. J Environ Eng 2005; 13: 396-402.

[52]   García-Peña I, Ortiz I, Hernández S, Revah S. Biofiltration of BTEX by the fungus *Paecilomyces variotii*. Int Biodeterior Biodegrad 2008; 62: 442-7.

[53]   Moe WM, Qi B. Performance of a fungal biofilter treating gas-phase solvent mixtures during intermittent loading. Water Res 2004; 38: 2259-68.

[54]   Spigno G, De Faveri DM. Modeling of a vapor-phase fungi bioreactor for the abatement of hexane: fluid dynamics and kinetic aspects. Biotechnol Bioeng 2005; 89(3):319-28.

[55]   Yaomin J, Veiga MC, Kennes C. Performance optimization of the fungal biodegradation of α-pinene in gas-phase biofilter. Process Biochem 2006; 41(8): 1722-8.

[56]   Arriaga S, Revah S. Removal of n-hexane by *Fusarium solani* with a gas-phase biofilter. J Ind Microbiol Biotechnol 2005; 32: 548-53.

[57]   Vergara-Fernández A, Hernández S, Revah S. Phenomenological Model of fungal biofilters for the abatement of hydrophobic VOCs. Biotechnol Bioeng 2008; (6): 1182-92.

[58]   Magaña-Reyes M, Morales M, Revah S. Methyl tert-butyl ether and tert-butyl alcohol degradation by *Fusarium solani*. Biotechnol Lett 2005; 27: 1797-801.

[59]   Vigueras G, Arriaga S, Shirai K, Morales M, Revah S. Hydrophobic response of the fungus *Rhinocladiella similis* in the biofiltration with volatile organic compounds with different polarity. Biotechnol Lett 2009; 31: 1203-9.

[60]   Vigueras G, Shirai K, Martins D, Teixeira-Franco T, Fleuri L, Revah S. Toluene gas phase biofiltration by *Paecilomyces lilacinus* and isolation and identification of a hydrophobin protein produced thereof. Appl Microbiol Biotechnol 2008; 80(1): 147-54.

[61]   Christen P, Domenech F, Michelena G, Auria R, Revah S. Biofiltration of Volatile Ethanol Using Sugar Cane Bagasse Inoculated with *Candida utilis*. J Hazard Mater 2002; 89, 253-65.

[62]   Ortíz I, García-Peña I, Christen P, Revah S. Effects of Inoculum type, packing material and operation conditions on pentane biofiltration. Chem Biochem Eng 2008; 22(2): 179-84.

[63]   Deshusses MA, Hamer G, Dunn IJ. Behavior of biofilters for waste air biotreatment. 2. Experimental evaluation of a dynamic model. Environ Sci Technol 1995; 29(4): 1059-68.

[64]   Kasenski S, Kinney KA. Biofiltration of Paint Spray Booth Emissions: Packing Media Considerations and VOC Interactions. In *the Proceedings of the 93rd Annual Meeting and Exhibition of the Air & Waste Management Association*, Salt Lake City, Utah 2000.

[65]   Atoche JC, Moe WM. Treatment of MEK and toluene mixtures in biofilters: effect of operating strategy on performance during transient loading. Biotechnol Bioeng 2004; 86(4): 468-81.

[66]   Auria R, Morales M, Villegas E, Revah S. Influence of mold growth on the pressure drop in aerated solid state fermentors. Biotechnol Bioeng 1993; 41(11): 1007-1013.

[67]   van Groenestijn JW, Kraakman NJR. Recent developments in biological waste gas purification in Europe. Chem Eng J 2005; 113: 85-91.

[68]   Ottengraf SPP, Konings JHG. Emission of microorganisms from biofilters. Bioprocess Biosyst Eng 1991; 7: 89-96.

[69]   Pineda J, Auria R, Pérez-Guevara F, Revah S. Biofiltration of toluene vapors using a model support. Bioprocess Biosyst Eng 2000; 23: 479-86.

[70]   Alonso C, Zhu X, Makram S. Parameter estimation in biofilter systems. Environ Sci Technol 2000; 34: 2318-23.

[71]   Bibeau L, Kiared K, Brzenzinski R, Viel G, Heitz M. Treatment of air polluted with xylenes using a biofilter reactor. Water Air Soil Pollut 2000; 118: 377-93.

[72]   Nopharatana M, Howes T, Mitchell DA. Modeling fungal growth on surfaces. Biotechnol Tech 1998; 12: 313-8.

[73]   Mitchell DA, von Meien OF, Krieger N, Dalsenter FDH. A review of recent developments in modeling of microbial growth kinetics and intraparticle phenomena in solid-state fermentation. Biochem Eng J 2004; 17: 15-26.

[74]   Arriaga S, Revah S. Mathematical modeling and simulation of hexane degradation in fungal and bacterial biofilters: Effective diffusivity and partition aspects. Can J Civil Eng 2009; 36(12):1919-25.

Index

A

abatement, 109, 118, 120
abiotic, 72
absorber, 110
absorption, 20, 47, 58, 109-110, 117
absortion, 62-63
ABTS, 64-66
acarbose, 15, 25
acetaldehyde, 116
acetate, 87, 100, 106, 114-116
acetic, 32, 87, 116
acetone, 113
acetosyringone, 36
acetovanillone, 36
acetyl, 32-34, 49, 52-53
acetylcholine, 99, 105
acetylgalactoglucomannan, 32
acetylglucomannan, 32-33
acetylxylan, 32-33
Achromobacter, 72
acidity, 85, 104
acids, 3, 9, 14, 23, 29, 32, 35-36, 39, 44-46, 57, 73-74, 81-89, 97-98, 106
Actinomucor, 74-75
actinomycetes, 24, 111
acylhydrolases, 82
additive, 20, 22, 74, 105
adduct, 65
adherence, 40
adhesion, 100, 106
adhesives, 6, 40
adipocyte, 100
adipogenesis, 106
adipogenic, 106
adsorption, 109
aeration, 104
affinity, 7, 15, 25, 54, 57, 64, 70, 117
aflatoxin, 70, 72, 93
Agaricus, 33, 66, 69-70, 72-73
agglomeration, 104
agitation, 104
agricultural, 4, 28, 36-38, 41
Agrobactrium, 105, 108
albocoremium, 89, 98
albomyces, 33, 60-63, 65-66, 70
Alcaligenes, 69, 71
alcohol, 6, 14, 29, 34-35, 39, 41-42, 82-83, 104, 115, 120
aldehyde, 3, 105
algae, 32, 76
algorythm, 62, 67
alignment, 60-62, 67, 69, 72

G

I

N

www.ingramcontent.com/pod-product-compliance
Lightning Source LLC
Chambersburg PA
CBHW041712210326
41598CB00007B/619